T0336560

Molecular Aspects of Aging

Molecular Aspects of Aging

Understanding Lung Aging

Edited by

Mauricio Rojas
Dorothy P. and Richard P. Simmons Center
for Interstitial Lung Diseases;
Division of Pulmonary, Allergy and Critical
Care Medicine;
McGowan Institute for Regenerative Medicine,
University of Pittsburgh School of Medicine,
Pittsburgh, Pennsylvania, USA

Silke Meiners
Comprehensive Pneumology Center (CPC),
University Hospital, Ludwig-Maximilians-University,
Helmholtz Zentrum München;
Member of the German Center for Lung Research (DZL),
Munich, Germany

Claude Jourdan Le Saux
University of Texas Health Science Center, Division of
Cardiology and Pulmonary and Critical Care, San Antonio,
Texas, USA

WILEY Blackwell

Library of Congress Cataloging-in-Publication Data

Molecular aspects of aging : understanding lung aging / edited by Mauricio Rojas, Silke Meiners, Claude Jourdan Le Saux.
    p. ; cm.
  Includes bibliographical references and index.
  ISBN 978-1-118-39624-7 (cloth)
  I. Rojas, Mauricio, 1963- editor of compilation.   II. Meiners, Silke, editor of compilation.
III. Le Saux, Claude Jourdan, editor of compilation.
  [DNLM: 1. Aging–physiology. 2. Lung–physiology. 3. Age Factors. 4. Lung Diseases–metabolism.
5. Lung Diseases–physiopathology.   WF 600]
  QP1837.3.A34
  612.6′7–dc23

                                                                                              2014000056

Set in 10/12pt Times by SPi Publisher Services, Pondicherry, India
Printed and bound in Malaysia by Vivar Printing Sdn Bhd

1   2014

Contents

Contributors

Serge Adnot
INSERM U955 and Département
de Physiologie-Explorations
Fonctionnelles, Hôpital Henri Mondor,
Université Paris Est, Paris, France

David E. Bloom
Department of Global Health and
Population, Harvard School of Public
Health, Boston, Massachusetts, USA

Jorge Boczkowski
INSERM U955 and Département de
Physiologie-Explorations Fonctionnelles,
Hôpital Henri Mondor, Université Paris Est,
Paris, France

Laurent Boyer
INSERM U955 and Département de
Physiologie-Explorations Fonctionnelles,
Hôpital Henri Mondor, Université Paris Est,
Paris, France

Rodrigo T. Calado
University of São Paulo at Ribeirão,
Preto Medical School, São Paulo, Brazil

Leena P. Desai
Division of Pulmonary, Allergy, and
Critical Care Medicine, Department
of Medicine, University of Alabama
Birmingham, Birmingham,
Alabama, USA

Deepak A. Deshpande
Pulmonary and Critical Care Medicine
Division, University of Maryland
School of Medicine, Baltimore,
Maryland, USA

George A. Garinis
Institute of Molecular Biology and
Biotechnology, Foundation for Research
and Technology-Hellas and Department of
Biology, University of Crete, Crete, Greece

Francis H.Y. Green
University of Calgary, Calgary, Alberta, Canada

Kevin P. High
Section on Infectious Diseases, Wake
Forest School of Medicine, Winston-Salem,
North Carolina, USA

Anne Hilgendorff
Comprehensive Pneumology Center (CPC),
University Hospital, Ludwig-Maximilians
University, Helmholtz Zentrum München;
Member of the German Center for Lung
Research (DZL); Dr. von Haunersches
Children's Hospital, Munich, Germany

Anna Ioannidou
Institute of Molecular Biology and
Biotechnology, Foundation for Research
and Technology-Hellas and Department of
Biology, University of Crete, Crete, Greece

Maria G. Kapetanaki
Dorothy P. and Richard P. Simmons Center
for Interstitial Lung Diseases, Division of
Pulmonary, Allergy and Critical Care
Medicine, University of Pittsburgh School
of Medicine, Pittsburgh, Pennsylvania, USA

Ismene Karakasilioti
Institute of Molecular Biology and
Biotechnology, Foundation for Research
and Technology-Hellas and Department of
Biology, University of Crete, Crete, Greece

Jacqueline M. Kruser
Department of Medicine, University of
Wisconsin School of Medicine and Public
Health, Madison, Wisconsin, USA

Claude Jourdan Le Saux
University of Texas Health Science Center,
Division of Cardiology and Pulmonary and
Critical Care, San Antonio, Texas, USA

Silke Meiners
Comprehensive Pneumology Center (CPC),
University Hospital, Ludwig-Maximilians-
University, Helmholtz Zentrum München;
Member of the German Center for Lung
Research (DZL), Munich, Germany

Keith C. Meyer
Department of Medicine, University of
Wisconsin School of Medicine and Public
Health, Madison, Wisconsin, USA

Ana L. Mora
Division of Pulmonary, Allergy and Critical
Care Medicine, University of Pittsburgh
School of Medicine, Pittsburgh,
Pennsylvania, USA

Kent E. Pinkerton
Center for Health and the Environment,
University of California Davis,
Davis, California, USA

Mauricio Rojas
Dorothy P. and Richard P. Simmons Center
for Interstitial Lung Diseases; Division of
Pulmonary, Allergy and Critical Care
Medicine; McGowan Institute for
Regenerative Medicine, University of
Pittsburgh School of Medicine, Pittsburgh,
Pennsylvania, USA

Jesse Roman
Department of Medicine, Division of
Pulmonary, Critical Care and Sleep
Disorders, Department of Pharmacology &
Toxicology, Robley Rex Veterans
Affairs Medical Center and University
of Louisville, Kentucky, USA

Yan Y. Sanders
Division of Pulmonary, Allergy, and
Critical Care Medicine, Department of
Medicine, University of Alabama
Birmingham, Birmingham, Alabama, USA

Sinead Shannon
Department of Health and Children,
Dublin, Ireland

Pooja Shivshankar
University of Texas Health Science Center,
Division of Cardiology and Pulmonary and
Critical Care, San Antonio, Texas, USA

Suzette M. Smiley-Jewell
Center for Health and the Environment,
University of California Davis,
Davis, California, USA

Victor J. Thannickal
Division of Pulmonary, Allergy, and
Critical Care Medicine, Department of
Medicine, University of Alabama
Birmingham, Birmingham, Alabama, USA

Lei Wang
Center for Health and the Environment,
University of California Davis,
Davis, California, USA

Mingyi Wang
Intramural Research Program, National
Institute on Aging, Baltimore, Maryland,
USA

Jingyi Xu
Center for Health and the Environment,
University of California Davis,
Davis, California, USA; Affiliated
Zhongshon Hospital of Dalian
University, Dalian, China

Preface

Aging is the inevitable fate of life. It is a natural process characterized by progressive functional impairment and reduced capacity to respond adaptively to environmental stimuli. The aging process, among other factors, determines the life span of an organism, whereas age-associated abnormalities account for the health status of a given individual. Aging is associated with increased susceptibility to a variety of chronic diseases, including type 2 diabetes mellitus, cancer, and neurological diseases. Lung pathologies are no exception, and the incidence and prevalence of chronic lung diseases has been found to increase considerably with age.

Aging has various faces, and most importantly, it has no purpose. Age-related pathologies are believed to result from the accumulation of molecular and cellular damage that cannot be repaired by aged cells due to limited performance of somatic maintenance and repair mechanisms. Two major hypotheses provide a conceptual framework for aging. According to the **antagonistic pleiotropy** hypothesis by Williams (Evolution, 1957), natural selection favors genes that are beneficial early in life for the cost that they may promote aging later in life. The **disposable soma** theory put forward by Kirkwood (Nature, 1977) proposes that the organism optimally allocates its metabolic resources, chiefly energy, to maximize reproduction, fitness, and survival. This comes at the cost of limited resources for somatic maintenance and repair causing accumulation of molecular and cellular damage. This concept supports the observation that the aging process is stochastic in nature and that there is individual plasticity.

The objectives of this book are to increase our awareness and knowledge of the physiological and accelerated mechanisms of the aging lungs given the expected increase in the aging population in the coming years. We would like to stimulate research on the molecular aspects of lung aging by combining chapters on the general hallmarks of aging with chapters on how to analyze lung aging by experimental approaches and chapters on the molecular and clinical knowledge on physiological and premature aging in lung disease.

As outlined in Chapter 1, the aging population will be more and more vulnerable to developing pathological conditions due to age-associated morbidities. Chapters 2–7 summarize characteristic cell-autonomous and systemic hallmarks of aging. While Chapter 2 gives an overview on the transcriptomic signatures of the aging organism, Chapters 3 and 4 introduce loss of proteostasis and the molecular details of telomere dysfunction, respectively. In Chapters 5 and 6, cellular senescence – the cell-autonomous aging program – is outlined in detail, and cellular signaling pathways that control senescence are elucidated. Chapter 7 provides an overview on the age-related changes of the immune system. Chapters 8–14 focus on the aging lung and age-related pathologies of the lung. Chapter 8 introduces the physiological aging process of the lung which is characterized by senile lung emphysema and the age-related decline in lung function in the elderly. Mouse models to explore the molecular nature of age-related lung pathologies are summarized in Chapter 9. Early damage of the immature lung as observed in neonates contributes to premature lung aging as outlined in Chapter 10. The aging lungs present featured changes of the extracellular matrix (Chapter 11) and of the mesenchymal stem cell compartment (Chapter 12). While age-related changes in tissue repair such as altered matrix remodeling and stem cell recruitment add to fibrotic pulmonary diseases, telomere dysfunction and cellular senescence are hallmarks of premature aging in chronic obstructive pulmonary disease (Chapter 13). Immunosenescence and inflamm-aging both promote impaired host responses to respiratory infections in the elderly as outlined in Chapter 14.

We hope that this book will attract basic and clinical scientists to study the mechanisms of aging in general and of the lung in particular. We are confident that the book will contribute to our understanding of age-related lung diseases, and we wish you pleasure reading this book!

1 The Demography of Aging

David E. Bloom[1] and Sinead Shannon[2]

[1] Department of Global Health and Population, Harvard School of Public Health, Boston, Massachusetts, USA
[2] Department of Health and Children, Dublin, Ireland

1.1 Introduction

Throughout the world, people are living longer, healthier lives, and the proportion of older people is growing more rapidly than ever before, causing a dramatic shift in the global population age structure. These trends have been clear for several decades, and with each passing year, research reveals more about how the changing demographic structure is likely to affect individuals and societies. The ongoing changes will have implications for the development of policy in a number of areas – such as health, pensions, education, finance, and job structures – and although population aging is frequently presented as a threat, mitigating factors can dramatically alter its impact.

This chapter examines current age profiles throughout the different regions of the world. It starts with the factors contributing to the growth in the absolute numbers and proportion of older people and then looks at the factors contributing to the potential economic and health impact of aging. On the health front, the challenge will be to balance longer lives with an increase in the number of healthy years. If this can be done, it will help society control outlays on health and social care, along with enabling older people to live more productive, fulfilling lives.

1.2 Demographic trends

Let us start with the sequence of demographic changes known as the **demographic transition** – which all countries experience at varying paces and to varying degrees as they evolve from agrarian societies to modern industrial ones. This transition has four phases: (i) pretransition equilibrium at high levels of both mortality and fertility; (ii) mortality declines and fertility remains high, leading to a growth in the size of the population; (iii) population growth reaches its peak, followed by a decline in the crude birthrate that is faster than the decline in the crude death rate, leading to a slowing of population growth; and (iv) posttransition equilibrium at low levels of mortality and fertility [1].

Molecular Aspects of Aging: Understanding Lung Aging, First Edition. Edited by Mauricio Rojas, Silke Meiners and Claude Jourdan Le Saux.

In Europe and North America, the first phase took place during the several hundred years prior to the Industrial Revolution. It was typified by a high birthrate and a death rate that fluctuated because of epidemics and famines. After the Industrial Revolution, European countries started to see a decline in the mortality rate as public health improvements began to have an effect. In the two decades following World War II, the birthrate rose initially but then gradually declined throughout the remainder of the 20th century while the mortality rate also fell. The current stage of the developed world's demographic transition is characterized by a birthrate at replacement level (roughly 2.1) or below in many countries and a steady increase in longevity.

1.2.1 Fertility rates

In developed countries, fertility rates have been falling for a number of decades and reached replacement level around 1975 [2]. The European Union (EU) experienced a sharp fall in fertility rates between 1980 and the early 2000s, reaching 1.47 in 2003. However, since 2005, there has been an increase in almost all countries in the EU-27, resulting in an average of 1.59 in 2009 [3]. In the United States, the rate now stands at 2.1 children – the long-run replacement rate [2]. While the global fertility rate can vary dramatically, UN figures show that the number of countries with high fertility has gradually declined and is projected to continue falling. In 2000–2005, 56 countries (out of 192) had a total fertility of 4.0 or higher but by 2045–2050, the fertility rate, even among what are today's developing countries, is projected to fall to roughly 2.2 (and to about 2.8 in the least developed countries) [2].

1.2.2 Mortality rates and life expectancy

In the past, increases in life expectancy stemmed disproportionately from reductions in child mortality rates, but in the future, the UN predicts that the impetus will increasingly come from a reduction in mortality at the older ages.

Back in 1700, life expectancy at birth in England, which was at the time one of the richest countries in the world, was only 37 years [4]. The development of antibiotics and vaccines and subsequent improvements in hygiene, sanitation, and public health led to reductions in mortality at all ages, especially in childhood. More recently, as countries became more prosperous, economic development contributed to improved nutrition, immunization against common diseases, and a consequent reduction in death rates worldwide. It is thought that the introduction of clean water and improved sanitation in the United States during the late 19th and early 20th centuries may have been responsible for reducing mortality rates by about half and child mortality rates by nearly two-thirds in major cities. In the country overall, the death rate fell by 40% – an average decline of about 1% per year [5].

Globally, infant mortality has fallen from 51 deaths per 1000 in 2000 to 42 in 2010, with the rate projected to decline to 23 per 1000 by 2050 [2]. In OECD countries, infant mortality rates have seen a dramatic reduction from a level of 41 deaths per 1000 births in 1970 to an average of 8 deaths in 2010. However, substantial variations occur within countries. In the United States, for example, the infant mortality rate for children born to African-American mothers is more than double that for white women (12.9 vs. 5.6 in 2006) [6, 7].

Life expectancy at birth varies greatly across countries and levels of development, from as low as 57 years in less developed regions (2005–2010) to 77 years in more developed regions. Although, globally, it is predicted to increase to 69 years by 2050, this will depend largely on succeeding in the fight against HIV/AIDS and other infectious diseases. In the EU-27, life expectancy for men in 2009 ranged from 67.5 years in Lithuania to 79.4 in Sweden [8].

1.2.3 Proportion of older people

The effect of increasing life expectancies and low levels of fertility, sustained for decades, has been an overall increase in the proportion of older people, accompanied by a lower proportion of younger people. In the United States, partly as a result of lower fertility, the population is growing slowly and beginning to age rapidly. For example, between 2010 and 2011, the number of young people (aged under 20) increased by only 375,000 (0.4%), while the number of people aged 60 and older (the 60+) increased by 1.58 million (2.8%) [2].

The UN estimates that, globally, the proportion over 60 will increase from 11% to 22% by 2050 (see Figure 1.1) – and will reach 28% by 2100. Although the world's population will triple in size by 2050 (from 1950), the number of people who are 60+ is expected to increase by a factor of 10, and those 80+ by a factor of 27 [2].

The proportion of people aged 60+ is not only changing over time but also varies greatly by region. Among countries, Japan currently has the largest proportion of people (30%) aged 60+ – a title that it is expected to still hold in 2050 when the figure reaches 44%. By then, every country in the world is expected to have a higher 60+ share, at which time one-third of the world's population will be living in countries with a higher proportion of older people than Japan has now [2]. Currently, the US share of 60+ is 18% (57 million), which is expected to rise to 27% (107 million) in 2050 – and to 31% (149 million) by 2100.

Similar trends appear in the population aged 80+, with Africa the only region not projected to have a very rapid increase in the proportion of the population in this age group. In the United States, the share of those aged 80+ is predicted to rise to 8% (32 million) by 2050, up

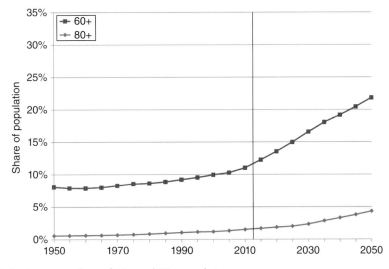

Figure 1.1 Increasing share of 60+ and 80+ population.

from 4% (12 million) in 2010. By 2050, the number of US centenarians is projected to reach nearly half a million [2]. The impact of all these demographic changes is that:

- The number of people aged 60+ will overtake the number of children (those aged 0–14) by 2047 [2].
- The bulk of population growth is expected to come from the developing world – with Africa's population projected to rise from 1 billion in 2010 to 3.6 billion in 2100.
- By 2100, only about 13% of the world's population will live in today's rich countries, down from 32% in 1950.

Of course, the rate of change in the share of older people in the population will vary from country to country, but less developed regions as a whole will experience a more rapid pace of growth. The countries that are expected to age most rapidly between 2010 and 2050 are primarily in the Middle East and Asia, while the least rapidly aging countries are in Africa [2].

1.3 Impact of aging

Over the past few decades, as population aging has become an issue for governments, particularly in the developed world, there has been considerable debate about whether we can expect the additional years of life to be healthy and active ones and whether governments will be able to meet the social and economic challenges that aging brings.

1.3.1 Noncommunicable disease trends

The prevalence of chronic conditions – known as noncommunicable diseases (NCDs) – has risen during the past two decades in both developed and developing countries [9]. Success in reducing communicable diseases has led to the dramatic increase in life expectancy (particularly in developed countries). However, the concern now is that the increase in the prevalence of NCDs will lead to a growth in the aggregate disease burden (because NCDs, by their nature, are generally long-lasting) and ultimately result in unsustainably high health costs (because they are typically expensive to treat). Moreover, any increased health costs may need to be paid by a relatively smaller working-age population because of the changing demographic structure.

The World Health Organization (WHO) estimates that in 2008, the four main NCDs (cardiovascular disease (CVD), cancer, chronic respiratory diseases, and diabetes) were responsible for the deaths of more than 31 million people worldwide – with about one-fourth of all NCD deaths premature (under 60 years) [9].

In the United States, it has been estimated that 65% of all health-care spending is on people with at least one chronic condition and that two-thirds of all Medicare spending is on people with five or more conditions [10]. This in itself is not necessarily a negative statement; rather, it is a reflection of the reduction in other causes of death such as communicable diseases and accidents. The prevalence of NCDs is rising in less developed regions, with roughly 80% of all NCD deaths now occurring in low- and middle-income countries [9]. Of note, NCD mortality appears to be more premature in low- and middle-income countries than in high-income countries. This presumably reflects the fact that in poorer countries risk factors are more prevalent and there is less prevention, early detection, and access to treatment.

Although deaths from heart disease, cancer, and stroke are declining, those from chronic obstructive pulmonary disease (COPD) are growing. Lung problems – such as COPD, chronic bronchitis, emphysema, asthma, and airflow obstruction – are among the key contributors to the global burden of disease. COPD was the sixth leading cause of death worldwide in 1990 and is expected to become the third by 2020. According to WHO estimates, 235 million people currently have asthma [11] and 65 million people have COPD [12]. Among cancer deaths, lung cancer is the biggest killer throughout the world [9].

How prevalent is COPD in the United States? The figure varies between 3% and 10%, depending on the diagnostic criteria. The Centers for Disease Control and Prevention puts the figure at 6.3% of adults [13] and estimates that the disease is responsible for about 700,000 hospitalizations and more than 130,000 deaths in 2009. By 2020, it is expected to be responsible for the deaths of more individuals than stroke. Moreover, there is evidence that COPD may be underdiagnosed and undertreated in older people [14].

1.3.2 Risk factors

NCDs stem from a combination of modifiable and nonmodifiable risk factors. The latter refers to characteristics that cannot be changed by an individual (or the environment), such as age, sex, and genetic makeup. The former refers to characteristics that societies or individuals can change to improve health outcomes – primarily (i) poor diet, (ii) physical inactivity, (iii) tobacco use, and (iv) harmful alcohol use. The pathway from modifiable risk factors to NCDs often operates through what are known as **intermediate risk factors** – which include overweight/obesity, elevated blood glucose, high blood pressure, and high cholesterol [15]. Environmental toxins present in the air and water and on land also appear to play a role.

The WHO Global Status Report on NCDs (2010) found that the key underlying causes of death globally from NCDs are raised blood pressure (responsible for 13% of deaths), tobacco use (9%), raised blood glucose (6%), physical inactivity (6%), and overweight/obesity (5%) [9]. It also estimated that smoking causes about 71% of all lung cancer deaths and 42% of chronic respiratory disease [16]. The prevalence of these risk factors varies among regions and by gender and income level. In high-income countries physical inactivity among women, total fat consumption, and raised total cholesterol were the biggest risk factors, whereas in middle-income countries tobacco use among men and overweight and obesity were the biggest contributors to NCDs. The prevalence of smoking is higher in middle-income countries than in low- or high-income countries, and in all income groups, higher among men than women. Of the six WHO regions, the highest overall prevalence for smoking in 2008 was estimated to be the European region, at nearly 29% [9].

People in high-income countries are more than twice as likely to get insufficient exercise, with 41% of men and 48% of women insufficiently physically active, compared with 18% of men and 21% of women in low-income countries [9].

Obesity is more prevalent in high-income countries, where more than half of all adults are overweight and just over one-fifth are obese. However, overweight/obesity has recently started to affect lower-income countries, with the increase in prevalence from 1980 to 2008 (a doubling) greater than in upper-middle- and high-income countries [17].

In developing countries, the increase in NCDs can be attributed to factors less common in developed countries, such as malnutrition in the first 1000 days of life and environmental pollution – even though the major risk factors are also common in developing countries [18].

Within countries, the difference in life expectancy between the highest and lowest socioeconomic groups is also increasing, reflecting the greater prevalence of NCDs at younger ages in lower socioeconomic groups. In addition, mortality from NCDs shows a threefold difference between the highest and lowest occupational classes in some countries – perhaps reflecting in part that those with lower education levels are less likely to receive a medical diagnosis, and even after diagnosis, experience greater difficulty in managing their conditions, or adhering to a disease management program. The evidence also suggests that it is education level rather than income level that has the greatest impact on health [19].

1.3.3 Impact of NCDs on health and disability

One of the key issues in relation to the rise in NCDs is whether they result in a burden, either for the individual or for the state. Living with an NCD for many years may not represent a major burden to the individual or his/her family unless the disease results in disability or infirmity and prevents them from continuing to work. However, lengthy periods of ill-health or disability will raise the cost of providing health services to the increasing numbers of older people. As all individuals must die from some cause, the aim must be to reduce the impact of the disease and minimize any reduction in health-related quality of life for the individual – and if possible, compress the period of illness or disability into a shorter period of time.

Research has not fully clarified how NCD trends are linked to the prevalence of disability and whether an extension of life expectancy will result in additional healthy years or an expansion of morbidity. Differing theories have been put forward since Gruenberg [20] predicted a pandemic of chronic diseases or expansion of morbidity. In 1980, James Fries [21] suggested that instead we would see a **compression of morbidity** – that is, the postponement of disease and disability and the compression of ill-health, activity limitation, or disability into a shorter period of time prior to death. In 1982, Manton [22] proposed a middle-ground theory that argued in favor of the emergence of a **dynamic equilibrium**, where the prevalence of disability would increase as mortality falls but the severity of disability would decline.

Who is right? The evidence supporting each of these theories is mixed, partly because of differing definitions of disability. As for the compression of morbidity thesis, some studies are supportive. For example, one of them that compared two groups of over 50s over a period of 21 years found that those who undertook regular vigorous exercise (members of a running club) reached a particular level of disability 12 years later than those in the control group (7 years vs. 19 years) [23]. However, others argue that there is substantial evidence to suggest that progress toward the elimination or delay of disease linked to aging has been limited – for example, the incidence of a first heart attack has remained relatively stable between the 1960s and 1990s, and the incidence of some of the most important cancers has been increasing until recently [24].

With regard to the dynamic equilibrium theory, one particularly supportive study argues that while the prevalence of many diseases has increased, there has been a reduction in the impact of such diseases on the individual, being both less lethal and less disabling [24]. Differing suggestions have been put forward to explain the variations and apparent contradictions between countries and over time. Deeg suggests that the initial level of disability may influence findings – that is, countries with initially high levels of disability provide more potential for reduction and therefore compression, while others with low starting levels (based on the time from which data is available) offer less potential for reduction [25]. However, Robine and Jagger [26] suggest that the demographic and epidemiologic theories of population health transition provide the answer. They argue that countries, genders, and socioeconomic groups within countries may be at different stages of a general health transition: first, people survive various illnesses into older ages and disability rises; then, the health of older people improves through various means and the number of years lived with disability decreases; but finally, the number of years lived with disability rises again when the average age of death rises to the extent that many people spend their last years at advanced old age with multiple chronic diseases and frailty [24, 27].

1.3.4   Increase in multimorbidities

Many of the most common NCDs frequently occur with other conditions, and growing numbers of people have more than one condition, especially if they smoke. The presence and increasing prevalence of multiple and costly comorbidities, particularly in later life, suggests a clear need for a new approach to preventing and treating such conditions. Each condition can influence the care of the other conditions by limiting life expectancy and the ability to remain active. Multimorbidities can also lead to interactions between therapies, and often, the treatment of one condition can inhibit the treatment for or exacerbate another condition.

Risk factors such as tobacco smoking can increase the likelihood of a person having a number of chronic conditions simultaneously. Studies show that COPD, particularly among older people, is characterized by a high prevalence of comorbid conditions such as CVD, muscle wasting, and osteoporosis. Among patients with COPD, the prevalence of heart failure varies across studies between 7.2% and 20.9% (depending on the diagnostic criteria used in the research); among patients with heart failure, the number of people diagnosed with COPD varies from 10.0% to 39.0% [28]. Depression, anxiety, and malnutrition are also common among older COPD patients [29].

The fact that people with chronic illnesses may be treated for a number of conditions, often by different medical professionals, results in their being prescribed a number of different drugs at the same time. In some cases, these drugs may have adverse interactions. In Europe, up to one-third of people aged 65 years and older use five or more prescription medications [30]. Boyd et al. [31] showed how, by following existing clinical practice guidelines, a hypothetical 79-year-old woman with COPD, type 2 diabetes, osteoporosis, hypertension, and osteoarthritis would be prescribed 12 medications, a mixture that risks multiple adverse reactions among drugs and can lead to avoidable hospitalization. Indeed, this is a major concern—among older adults, up to 16% of hospital admissions are due to adverse drug reactions [32].

1.3.5   Impact on expenditure

There is growing concern that population aging will drive up health-care expenditures. This concern is consistent with the positive cross-country correlation observed between the rising share of gross domestic product (GDP) devoted to health-care expenditures and the rising share of population at the older ages. However, evidence from a number of countries suggests that the costs associated with intensive hospital use prior to death are lower at older ages.

In the United States, the difference between hospital costs for those who died aged 85 and older is estimated to be 50% lower than for those who died between 65 and 69 [33], while in Denmark these costs are estimated to be 70% lower [34]. Why is this so? A Canadian study that sought to explain these differences identified a drop in hospital costs by 30% to 35% between the oldest and youngest age groups but found that the number of days spent in the hospital varied little between the two age cohorts. What was significant was the intensity of services received in the hospital day, a finding supported by research carried out on U.S. Medicare costs [33, 34]. Other research suggests that the bulk of expenditure for most people is likely to be required during the last year or two of life, regardless of age [37].

1.4 Policy responses

The prevalence of NCDs tends to be linked to either age or lifestyle – and fortunately, lifestyle-related causes can be modified, mitigated, or prevented by early intervention. In fact, World Bank evidence suggests that more than half of the NCD burden could be avoided through effective health promotion and disease prevention programs that tackle the prevalence of risk factors and reduce the number of premature deaths attributable to NCDs [38].

1.4.1 Preventing and managing NCDs

These interventions can occur at different stages in life: primary prevention could take place throughout the life course, focusing on the modifiable risk factors – such as better nutrition, more physical activity, higher rates of immunization, and health literacy (especially on smoking and alcohol risks). Secondary prevention can be most relevant to people aged 40–50 by focusing on known risk indicators (such as blood pressure, cholesterol, and low bone mass) – perhaps by giving users of health services more information and guidelines for self-management. Tertiary prevention occurs when the disease is present; it includes better disease management and rehabilitation from COPD, stroke, etc.

Many countries are developing new policy frameworks to prevent the occurrence of chronic disease and to manage the diseases in a way that delays the onset of complications and reduces emergency hospital admissions and use of expensive acute services. These policies include (i) increased health information, screening, and health checks for particular age groups, as well as deterrents such as increased taxation for tobacco or alcohol; (ii) less marketing of particular foods and beverages to children; (iii) taxes on foods that are high in sugar, salt, or fat; and (iv) earlier diagnosis and better treatment.

There is ample evidence of the success of such policies and that even simple measures can contribute to a reduction of the level of premature death. The WHO and the NCD Alliance (an association of "four international NGO federations representing the four main NCDs – cardiovascular disease, diabetes, cancer, and chronic respiratory disease") estimate that primary prevention measures can prevent 80% of premature heart disease, 80% of type 2 diabetes, and 40% of all cancers. Similarly, there is some evidence that secondary prevention can lower service use by between 7% and 17% at a very low cost [38]. Earlier and better treatment initiatives have reduced the number of people with heart disease and improved survival after cardiovascular events, which, in turn, has lowered CVD deaths [39].

A case study in North Karelia, Finland, exemplifies how the preventive approach can succeed. During the 1960s, the region had one of the highest rates of death in the world from coronary heart disease (CHD), especially among men. Following a large-scale preventive program, involving local and national authorities, the media, NGOs, supermarkets, the food industry, agriculture, health services, schools, and WHO experts, the level of smoking had fallen dramatically, dietary habits had improved, and most significantly, the prevalence of CHD had decreased [40].

1.4.2 Promoting exercise

There is considerable evidence supporting the benefits of physical exercise in maintaining health and physical functioning as people age. Exercise increases strength and is associated with a lower incidence of CVD, osteoporosis and bone loss, and certain forms of cancer. It can reduce the risk of falls, stroke, and insulin sensitivity and lower blood pressure among those suffering from hypertension [41]. The Swedish National Institute of Public Health calls exercise the **best preventive medicine for old age**, significantly reducing the risk of dependency in old age [42].

Unfortunately, older adults do not appear to be reaping the full benefits of exercise. One UK study found that physical activity declined rapidly at around the age of 55 and a third of people over 55 do not exercise at all (compared with 10% of people aged 33–54). In practice, regardless of age, relatively few people are doing enough physical exercise to protect their health [43, 44].

1.4.3 Monitoring health-risk behaviors (and chronic health conditions)

This is essential to developing health promotion activities, intervention programs, and health policies. But creating effective programs is not straightforward. In the United Kingdom, population screening for all people over 75 years of age was dispensed after it was found that it resulted in little or no improvement to quality of life or health outcomes [45]. However, more specific types of screening show promise. One way is to focus on older people at the point of contact with services (known as opportunistic screening), followed by active management by the appropriate medical professionals [46].

Another is to screen for particular conditions. In 2008, Abu Dhabi launched a prevention program that carried out simple screening for cardiovascular risks on 95% of the population in its first few years. Following screening each person received an individual report, outlining their main risk areas (such as high blood pressure or high body mass index), along with a range of recommended actions (such as dietary and exercise changes, or being assessed by their general practitioner). A recent overall assessment shows that the project identified a significant level of undiagnosed conditions – up to one-third of people with diabetes, one-half with hypertension, and two-thirds with high cholesterol were undiagnosed. It also found that the program achieved a 40% improvement in blood glucose levels and a 45% improvement in lipids, at a very low cost – less than US$20 per person per year [47].

1.5 Conclusion

How aging will affect future expenditures will be determined by many factors relating to (i) the demand for health and care services; (ii) the number of people that are high users of health services (particularly acute services); (iii) the length of time that they remain in the category of high users; and (iv) the cost of the health services they use. The need for care, in turn, tends to be determined by the presence of disability or inability to perform the typical activities of daily life.

With increasing numbers of older people in the population, the simplistic view is that this trend will lead to more people using services and a consequent increase in expenditure. Compounding matters is the higher prevalence of NCDs and multi-morbidities, which suggest a greater need for complex treatments and medication and more assistance or care.

However, there are a number of alternative ways to view this situation.

- First, not all those with NCDs experience activity limitations or need help with activities. We know that instrumental activities of daily living (IADLs: shopping, managing money, doing laundry, preparing meal, or using the telephone) have become easier to perform over the past couple of decades thanks to major improvements in the built and technical environments [48]. For example, technological advances enable older people with mobility difficulties to shop online or receive care remotely.
- Second, we know that advances in biomedical research and clinical innovations have reduced the rate of progression of diseases such as CVD and cancer and lowered disability rates [49].
- Third, we know that policy measures that are targeted at improving diet, increasing levels of physical activity, and reducing risk factors such as smoking can lower the prevalence and impact of NCDs.

The bottom line is that the target for public health policy should be to balance the gains in life expectancy with an equivalent increase in healthy life expectancy. Bringing about a change in the numbers of people using services, the length of time they use those services, and the cost and intensity of the services needed should help control expenditures. In addition, achieving greater integration and greater efficiency in managing and delivering services to older people can bring about an improvement in the health and quality of life of older people.

References

1. Chesnais, J.-C. Demographic transition patterns and their impact on the age structure. *Population and Development Review*, 1990;16:327–336.
2. United Nations. World Population Prospects: The 2010 Revision, Volumes I & II: Demographic Profiles. ST/ESA/SER.A/317, 2011.
3. European Commission. The 2012 Ageing Report: Economic and Budgetary Projections for the EU-27 Member States (2010–2060). Joint Report prepared by the European Commission (DG ECFIN) and the Economic Policy Committee (Ageing Working Group), 2012.
4. Wrigley, E.A. and Schofield, R. The Population History of England, 1541–1871: A Reconstruction. Cambridge: Harvard University Press, 1981.
5. Cutler, D.M. and Miller, G. The role of public health improvements in health advances: the twentieth-century United States. *Demography*, 2005;42:1–22.
6. NCHS Understanding Racial and Ethnic Disparities in U.S. Infant Mortality Rates. NCHS Data Brief, 2011:74. http://www.cdc.gov/nchs/data/databriefs/db74.htm (accessed on November 4, 2013).
7. OECD. "Infant mortality", in Health at a Glance 2011. Paris: OECD Publishing, 2011. http://dx.doi.org/10.1787/health_glance-2011-10-en (accessed on November 4, 2013).
8. EUROSTAT. Mortality and Life Expectancy Statistics. http://epp.eurostat.ec.europa.eu/statistics_explained/index.php/Mortality_and_life_expectancy_statistics#Further_Eurostat_information (accessed October 10, 2012).
9. Alwan, A. Global Status Report on Noncommunicable Diseases 2010: Description of the Global Burden of NCDs, Their Risk Factors and Determinants. Geneva: World Health Organization, 2011.
10. Benjamin, R.M. Public Health Reports, 2010;125:626–629.
11. WHO. Asthma: Fact sheet No. 307. Geneva: World Health Organization, 2013. http://www.who.int/mediacentre/factsheets/fs307/en/index.html (accessed on January 10, 2014).
12. WHO. Global Alliance Against Chronic Respiratory Diseases Action Plan 2008–2013. Italy: World Health Organization, 2008.

13. Centers for Disease Control and Prevention (CDC). Deaths from chronic obstructive pulmonary disease—United States, 2000–2005. *MMWR Morbidity and Mortality Weekly Report*, 2008;57: 1229–1232.

14. Bhatt, N.Y. and Wood, K.L. What defines abnormal lung function in older adults with chronic obstructive pulmonary disease? *Drugs & Aging*, 2008;25:717–728.

15. Bloom, D.E., Cafiero, E.T., Jané-Llopis, E., et al. The Global Economic Burden of Non-Communicable Diseases. Geneva: World Economic Forum, 2011.

16. Lim, S.S., Vos, T., Flaxman, A.D., et al. A comparative risk assessment of burden of disease and injury attributable to 67 risk factors and risk factor clusters in 21 regions, 1990–2010: a systematic analysis for the Global Burden of Disease Study 2010. *The Lancet*, 2013;380:2224–2260.

17. Finucane, M., Stevens, G.A., Cowan, M.J., et al. National, regional, and global trends in body-mass index since 1980: systematic analysis of health examination surveys and epidemiological studies with 960 country-years and 9.1 million participants. *The Lancet*, 2011,377:557–567.

18. World Bank, Growing Danger of Non-communicable Diseases. Washington, DC: World Bank, 2011.

19. Smith, J.P. The impact of socioeconomic status on health over the life-course. *Journal of Human Resources*, 2007;42:739–764.

20. Gruenberg, E.M. The failures of success. *The Milbank Memorial Fund Quarterly: Health and Society*, 1977;55:3–24.

21. Fries, J.F. Aging, natural death, and the compression of morbidity. *New England Journal of Medicine*, 1980;303:130–135.

22. Manton, K.G. Changing concepts of morbidity and mortality in the elderly population. *The Milbank Memorial Fund Quarterly: Health and Society*, 1982;60:183–244.

23. Fries, J.F., Bruce, B., and Chakravarty, E. Compression of morbidity 1980–2011: a focused review of paradigms and progress. *Journal of Aging Research*, 2011;Article ID 261702:10 pages.

24. Crimmins, E.M. and Beltrán-Sánchez, H. Mortality and morbidity trends: is there compression of morbidity? *The Journals of Gerontology Series B: Psychological Sciences and Social Sciences*, 2010;66:75–86.

25. Deeg, D.J.H., Robine, J.M., and Michel, J.P. 'Looking forward to a general theory on population aging': population aging: the benefit of global versus local theory. *Journal of Gerontology: Medical Sciences*, 2004;59:600.

26. Robine, J.M. and Jagger, C. The relationship between increasing life expectancy and healthy life expectancy. *Ageing Horizons*, 2005;3:14–21.

27. Robine, J.M. and Michel, J.P. Looking forward to a general theory on population aging. *Journal of Gerontology: Medical Sciences*, 2004;59A:590–597.

28. Mascarenhas, J., Azevedo, A., and Bettencourt, P. Coexisting chronic obstructive pulmonary disease and heart failure: implications for treatment, course and mortality. *Current Opinion in Pulmonary Medicine*, 2010;16:106–111.

29. Global Initiative for Chronic Obstructive Lung Disease (GOLD). From the Global Strategy for the Diagnosis, Management and Prevention of COPD, 2011. http://www.goldcopd.org (accessed on November 4, 2013).

30. Junius-Walker, U., Theile, G., Hummers-Pradier, E. Prevalence and predictors of polypharmacy among older primary care patients in Germany. *Family Practice*, 2007;24:14–19.

31. Boyd, C.M., Darer, J., Boult, C., Fried, L.P., Boult, L., Wu, A.W. Clinical practice guidelines and quality of care for older patients with multiple comorbid diseases: implications for pay for performance. *JAMA: The Journal of the American Medical Association*, 2005; 294:716–724.

32. Oxley, H. Policies for Healthy Ageing: An Overview. Paris: OECD, 2009.

33. Yang, Z., Norton, E.C., and Stearns, S.C. Longevity and health care expenditures: the real reasons older people spend more. *The Journals of Gerontology Series B: Psychological Sciences and Social Sciences*, 2003;58:S2–S10.

34. Madsen, J., Serup-Hansen, N., Kragstrup, J., and Kristiansen, I.S. Ageing may have limited impact on future costs of primary care providers. *Scandinavian Journal of Primary Health Care*, 2002;20:169–173.

35. Levinsky, N.G., Yu, W., Ash, A., et al. Influence of age on medicare expenditures and medical care in the last year of life. *JAMA: The Journal of the American Medical Association*, 2001;286:1349–1355.

36. McGrail, K., Green, B., Barer, M.L., Evans, R.G., Hertzman, C., and Normand, C. Age, costs of acute and long-term care and proximity to death: evidence for 1987–88 and 1994–95 in British Columbia. *Age and Ageing*, 2000;29:249–253.

37. Seshamani, M. and Gray, A.M. A longitudinal study of the effects of age and time to death on hospital costs. *Journal of Health Economics*, 2004;23:217–235.
38. Nikolic, I., Stanciole, E., and Zaydman, M. "Chronic Emergency: Why NCDs Matter," World Bank Health, Nutrition and Population Discussion Paper, 2011.
39. Fries, J.F., Koop, C.E., Sokolov, J., Beadle, C.E., and Wright, D. Beyond health promotion: reducing need and demand for medical care. *Health Affairs*, 1998;17:70–84.
40. Puska, P. Successful prevention of non-communicable diseases: 25 year experience with North Karelia project in Finland. *Public Health Medicine*, 2002;4:5–7.
41. Callaghan, P. Exercise: a neglected intervention in mental health care? *Journal of Psychiatric and Mental Health Nursing*, 2004;11:476–483.
42. Swedish National Institute of Public Health (SNIPH). Healthy Ageing: A Challenge for Europe. Brussels, Stockholm: National Institute of Public Health, 2007. www.healthyageing.nu and http://www.fhi.se/PageFiles/4173/Healthy_ageing.pdf (accessed on November 4, 2013).
43. Sports Council and Health Education Authority. Allied Dunbar National Fitness Survey. London: Sports Council/HEA, 1992.
44. US Department of Health and Human Services (PHS). Physical Activity and Health. A Report of the Surgeon General (Executive Summary). Pittsburgh, PA: Superintendent of Documents, 1996.
45. Fletcher, A., Price, G., Ng, E., Stirling, S., Bulpitt, C., and Breeze, E. Population-based multidimensional assessment of older people in UK general practice: a cluster-randomised factorial trial. *The Lancet*, 2004;364:1667–1677.
46. Stuck, A.E., Elkuch, P., Dapp, U., Anders, J., Iliffe, S., and Swift, C.G. Feasibility and yield of a self-administered questionnaire for health risk appraisal in older people in three European countries. *Age and Ageing*, 2002;31:463–467.
47. The Economist Intelligence Unit. Never too early: tackling chronic disease to extend healthy life years, 2012. http://digitalresearch.eiu.com/extending-healthy-life-years/report (accessed on November 25, 2013).
48. Spillman, B.C. Changes in elderly disability rates and the implications for health care utilization and cost. *Milbank Quarterly*, 2004;82:157–194.
49. Manton, K.G. Recent declines in chronic disability in the elderly US population: risk factors and future dynamics. *Annual Review of Public Health*, 2008;29:91–113.

2 The Omics of Aging: Insights from Genomes upon Stress

Ismene Karakasilioti, Anna Ioannidou, and George A. Garinis

Institute of Molecular Biology and Biotechnology, Foundation for Research and Technology-Hellas and Department of Biology, University of Crete, Crete, Greece

2.1 Introduction

Aging is an inexorable homeostatic failure of complex but largely unknown etiology that leads to increased vulnerability to disease (e.g., cancer, diabetes atherosclerosis, neurodegeneration) with enormous consequences on the quality of individual lives and the overall cost to society [1]. Human efforts over the last centuries have succeeded in substantially lengthening lifespan, allowing aging to become a common feature of Western societies. It has been, however, significantly challenging to unravel the molecular basis of the processes that cause loss of bodily functions and degeneration of cells and tissues with advancing age. The discouraging complexity of the aging process, the noticeable lack of tools to study it, and a shortage of experimentally tractable model systems have greatly hindered any testable hypothesis-driven approaches to understand the molecular basis of aging, particularly in mammals. It is now widely accepted that aging is not caused by active gene programming but by evolved limitations in somatic maintenance, resulting in a buildup of stochastic damage accumulation [2]. However, an accumulating body of evidence suggests that aging is also subject to regulation by evolutionarily highly conserved molecular pathways [3, 4, 5]. Recent progress driven largely by the use of advanced genomics approaches in accelerated aging (progeroid) and long-lived animals appears to reconcile both theories: macromolecular damage drives the functional decline with aging; however, a battery of conserved, longevity assurance mechanisms may set the pace on how rapidly damage builds up and function is lost over time (Figure 2.1). For instance, upon stress or other unforeseen circumstances, cells adjust the activity of hormonal pathways, dampen their oxidative catabolism, or increase their antioxidant responses while enhancing DNA repair and triggering proinflammatory signaling cascades or autophagy. At present, this intricate array of biological processes appears to impinge on mammalian lifespan in ways that are largely unknown and vastly unexplored. Nevertheless, the recent unprecedented progress in our ability to interrogate the genome and its function has provided us with important insights to better understand longevity assurance mechanisms. Coupled to physiological and molecular biological assays,

Molecular Aspects of Aging: Understanding Lung Aging, First Edition. Edited by Mauricio Rojas, Silke Meiners and Claude Jourdan Le Saux.

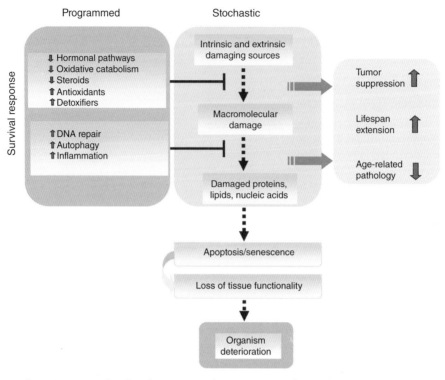

Figure 2.1 Programmed and stochastic events during aging. Stochastic macromolecular damage drives the functional decline with advancing age; however, evolutionarily conserved mechanisms may promote longevity by counteracting damage through, for example, DNA repair pathways or setting the pace on how rapidly damage builds up and function is lost through, for example, the regulation of hormonal pathways and antioxidant responses.

these unbiased **omics** approaches provide a solid basis for delineating the contributions from distinct and functionally diverse longevity-regulating networks in response to environmental or intrinsic stressors (e.g., DNA damage) during aging [2]. Understanding these connections will be instrumental for developing effective, rationalized intervention strategies against age-related cellular deterioration. Here, we focus on transcriptome analysis studies of a series of well-characterized, progeroid, and long-lived animal models.

2.2 Safeguarding the nuclear genome

In cells, the sources of macromolecular damage are broad, indiscriminately generating a range of chemical modifications to lipids, proteins, and DNA and unequivocally interfering with vital cellular processes, for example, transcription, translation, or replication, and metabolism. There is *a priori* no reason why certain types of damage would be more responsible in driving cellular deterioration over time than others. Likewise, there is no justification why a particular type of macromolecule would be more vulnerable to stochastic, indiscriminate damage than others are. However, unlike, for example, lipids or proteins that can be continually renewed or recycled, nuclear DNA, the blueprint of almost all cellular RNA and proteins, is truly unique in that it is neither **disposable nor recyclable** [6]. In spite of its massive length and inherent physicochemical instability, the somatic genome must be repaired once damaged and preserved during the entire lifetime of a cell. If not, any acquired error will be permanent, sparking off a plethora of irrevocable, life-threatening consequences. Immediate effects of DNA damage (i.e., the undesired chemical alteration in the base, sugar, or phosphate that alters the properties of DNA) include a physical block of transcription and replication. Long-term effects of DNA lesions involve induction of mutations (i.e., fixed errors in the coding sequence of an otherwise chemically unaffected DNA) via replication of damaged DNA, a major initiating and driving step for cancer. Eventually, the effects of distinct types of lesions diverge with respect to helix distortion, ability to pause or obstruct DNA replication and block ongoing transcription. Moreover, lesions are often **plastic** in nature; persistent, helix-distorting cyclobutane pyrimidine dimers (CPDs) induced by ultraviolet (UV) irradiation could give rise to DNA single- or double-strand breaks through damage-induced replication blockage [7]. It should, therefore, come as no surprise that DNA damage responses, including but not limited to DNA repair, frequently intersect with respect to lesion specificity [8].

Mammalian cells do more than meet the DNA damage challenge: cells have evolved elaborate machineries to maintain their chromosome ends (telomeres) intact, overlapping repair and genome surveillance pathways to counteract any structural DNA modifications, for example, nicks, gaps, double-strand breaks, and the myriad alterations that may block DNA transcription or replication [8, 9]. Together, these mechanisms warrant that the genetic information remains functionally intact for the entire natural life of a cell and is faithfully passed on to progeny. For bulky helix-distorting damage, such as the main UV-induced lesions, the principal repair mechanism is the evolutionarily conserved nucleotide excision repair (NER) pathway. NER recognizes and removes helical distortions throughout the genome (global genome NER, GGR) or selectively from the transcribed strand of active genes (transcription-coupled repair, TCR) [10]. In essence, GGR and TCR differ only in how helix-distorting DNA lesions are being recognized; following damage detection, the two subpathways then merge into a common mechanism to unwind the DNA around the lesion, stabilize and excise the DNA fragment containing the damage, and fill in and ligate the single-strand gap. In spite of the efficiency of genome mechanisms, however, DNA lesions gradually accumulate in mammals with age. This is mainly due to the fact that DNA repair mechanisms cannot cover all of the insults inflicted on the genome [11]. As the rate of DNA damage exceeds the capacity of the cell to repair it, so does the accumulation of errors that can result in loss of cellular homeostasis over time.

2.3 NER progerias and their connection to lifespan regulatory mechanisms

A number of inborn defects in DNA metabolism exist that are almost exclusively linked with an extending class of syndromes and associated mouse models with phenotypes resembling accelerated aging pointing to DNA damage as a major culprit in aging [12, 13, 14]. Patients with Cockayne syndrome (CS) carrying a NER defect (affected genes *CSB*, *CSA*) present with cachectic dwarfism and progressive neurological abnormalities (including mental retardation, microcephaly, gait ataxia, sensorineural hearing loss, retinal degeneration), along with impaired sexual development, kyphosis, osteoporosis, and severely reduced lifespan (mean age of death, 12.5 years) [15, 16]. Patients with trichothiodystrophy (TTD; affected genes: *XPB*, *XPD*, or *TTDA*) are partially defective in TCR, as well as in the GGR subpathway of NER, and share most of the symptoms associated with CS. In addition, these patients have a partial defect in transcription itself, causing additional symptoms such as ichthyosis and brittle hair and nails [17]. Many of the CS and TTD features are progressive and resemble premature aging. Similarly, mouse models carrying defects in NER develop prematurely a number of symptoms of aging (segments of aging) including progressive neurodevelopmental delay, cachexia, kyphosis, liver and kidney age-related pathology, and a substantially reduced lifespan that closely mimic those seen in corresponding patients with progeria [18, 19]. As patients and corresponding animal models develop some but not all aspects of normal aging in an accelerated manner, they are considered **segmental progeroid syndromes** [20].

Until recently, however, NER progeroid animal models had no functional link with longevity assurance mechanisms; their connection to aging was merely rationalized on the basis of the gradual accumulation of DNA lesions in the genome (due to the DNA repair defect) and the consequential premature onset of progeroid features leading to organismal deterioration over time. It was only recently that a series of transcriptome analysis studies in mice revealed that the rapid onset of age-related pathology in the DNA repair mutants (e.g., $Csb^{m/m}$, Xpd^{TTD}, $Csb^{m/m}/Xpa^{-/-}$, and $Xpd^{TTD/TTD}/Xpa^{-/-}$ and $Ercc1^{-/-}$) was associated with (i) a systemic suppression of the lifespan regulator growth hormone/insulin growth factor 1 (GH/IGF1) somatotrophic axis as well as of lactotroph and thyrotroph processes (i.e., the systemic response to prolactin and thyroid-stimulating hormones secreted by the pituitary); (ii) the suppression of oxidative metabolism (i.e., glycolysis, tricarboxylic acid cycle, and oxidative respiration); (iii) a reduction of serum glucose and insulin levels; and (iv) a consistent upregulation of antioxidant defense and stress responses ([19, 21, 22] and unpublished data on naturally aged $Csb^{m/m}$ mice from our lab). Furthermore, mice carrying defects in DNA damage signaling such as the Seckel syndrome mice with hypomorphic mutations in the ataxia telangiectasia and Rad3-related (ATR) protein kinase [23] or the ataxia telangiectasia-mutated (ATM) mutants [24], all of which display many symptoms of accelerated aging, tend to shift their expression towards growth suppression and upregulated antioxidant defenses. Similarly, mouse models for Hutchinson–Gilford progerias, which suffer from a defect in nuclear Lamin A, rendering nuclei particularly sensitive to, for example, mechanical stress, reveal gene expression changes that resemble those seen in NER progeroid animals [25]. This likely applies to the human situation as well, as suggested by the suppression of growth which is characteristic of many of the corresponding human progeroid repair syndromes, such as in CS, TTD, Seckel syndrome, ataxia telangiectasia, Hutchinson–Gilford progeria and variants, and XFE syndrome [26]. Thus, it seems that excessive cytotoxic DNA damage either caused by compromised repair or deficient damage signaling not only accelerates aging in the affected organs and tissues presumably by increased cell death and cellular senescence, but it is also intimately connected with metabolic and endocrine responses that cumulatively we call the **survival** response (Figure 2.1). Importantly, wild-type mice chronically exposed to subtoxic doses of oxidative damaging agents were previously shown to shift their gene expression profile towards suppression of the GH/IGF1 somatotrophic axis and upregulation of antioxidant systems, indicating that continuous (DNA) damage is able to substantially impinge on the GH/IGF1 somatotrophic axis [27].

2.4 Triggering a survival response in the absence of a DNA repair defect

The systemic dampening of the GH/IGF1 axis and energy metabolism together with the upregulation of antioxidant and detoxification defense genes are all associated with increased longevity as seen in tumor-resistant, long-lived pituitary mutant dwarfs and caloric-restricted (CR) mice rather than with the short lifespan of progeroid mice. Indeed, dietary restriction (DR) is found to trigger a similar highly interconnected set of homeostatic changes that impinge on numerous processes (Figure 2.2a). These include nutrient utilization and key energy-related metabolic pathways, growth, reproduction, resistance to (oxidative) stress, and for higher species inflammation and thermoregulation [28]. In addition, animals subjected to CR show improved insulin responses [29], reduced levels of glucose and low concentrations of advanced glycation end products resulting from the covalent modification of proteins by glucose derivatives, decreased levels of anabolic hormones (e.g., insulin, IGF1, thyroid hormones, testosterone, leptin), reduced body core temperature [30], as well as decreases in inflammatory cytokines (e.g., IL6, TNF-α) [31] and increases in anti-inflammatory hormones (e.g., adiponectin and ghrelin) [32, 33]. CR may also enhance various DNA repair systems, although systematic quantitative analyses are lacking [34]. In addition, it influences metabolism of xenobiotics (e.g., cytochrome P450 oxidases) and upregulates a battery of molecular and enzymatic antioxidant and detoxification systems (e.g., thioredoxins, superoxide dismutase, catalase, and peroxiredoxins) [35] to scavenge reactive oxygen species (ROS) and maintain biomolecules in their reduced state. This is expected to reduce accumulation of oxidative damage in DNA and other biomolecules, rendering the organism more resistant to oxidative stress [34]. Also, autophagy, the process by which cells recycle damaged organelles, has recently been found to be required for CR-mediated lifespan extension [36]. Together, such changes are thought to delay or avert age-related diseases, including atherosclerosis, diabetes, coronary heart disease, renal disease, and autoimmune and neurodegenerative diseases [30] (Figure 2.2a). Importantly, CR also confers significant protection against both spontaneous or radiation-induced neoplasias [37] and transplanted tumors in animal models [38], including nonhuman primates [39]. This effect may at least in part derive from the general antigrowth, promaintenance, and defense policy that CR imposes on the organism. Intriguingly, in several instances, these beneficial, prolongevity homeostatic changes seem to occur in the absence of any irreversible developmental or significant reproductive defects [28].

Besides CR, in mice, constitutive defects in single genes that perturb endocrine signaling can considerably extend lifespan. The most prominent pathway affected in endocrine-disturbed Ames and Snell dwarfs [40, 41], the little mouse (Ghrhr$^{lit/lit}$) [42], the growth hormone receptor/binding protein (Ghr/bp$^{-/-}$) [43], the heterozygous IGF1 receptor (Igf1r$^{+/-}$) knockout mice [44], or the Klotho-overexpressing mice [45] is the insulin/IGF-1 pathway. This is often paralleled by reduced glucose levels, suppressed thyrotroph and lactotroph functions, lower body temperature, diminished generation of free radicals, improved antioxidant defenses and stress resistance, enhanced storage of primary carbon sources, delayed development, reduced fertility, and increased average and maximal lifespan [46]. The great majority of these changes are also seen upon CR or with natural aging [19, 27, 47]. Similar to CR, one of the presumed mechanisms linking suppression of the IGF1/insulin signaling to longevity in dwarf rodent models is the consistent ability to protect against spontaneous tumors [46] or resist tumor induction following exposure to chemical carcinogens [48]. Conversely, overexpression of GH in GH-transgenic mice causes severe kidney pathology, a higher frequency of liver adenomas and carcinomas, fibrotic alterations, and a markedly shortened lifespan [49]. In line, injection of high GH doses in rats results in hepatomegaly and a 20% higher mortality rate [50]. Increasing IGF1/insulin signaling also appears to dampen the (antioxidant) stress responses in GH-transgenic mice [51] or long-lived Ames dwarf mice [52]. The fact that CR could induce a similar response to that seen in long-lived mutant

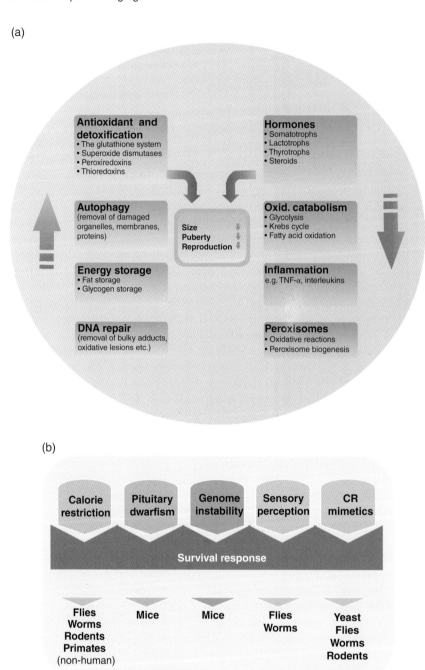

Figure 2.2 Components and instigators of the survival response in mammals. (a) The survival response represents a preservative metabolic strategy that aims at removing free radicals, toxins, damaged organelles, and membranes; delays age-related genomic decay and in the meantime suppresses growth and reproduction; and decreases inflammatory responses. In mammals, the various components of this conserved response result in smaller body size, delayed puberty, and decreased reproduction. (b) Shown are the major instigators that are presently known to trigger a generalized evolutionary survival strategy against a multitude of adverse physiological threats. However, there might be more cues involved in triggering the battery of physiological responses favoring longevity. To see a color version of this figure, see Plate 2.1.

dwarfs even in mice that are briefly exposed to CR regimens at later stages in life [53] suggests that longevity assurance mechanisms can be adaptive and reversible. Why are such longevity assurance responses previously only associated with the long-lived mutant dwarf and CR animals also seen in progeroid animals carrying defects in genome maintenance pathways? And given that there is sufficient evidence for both, is there a link between the genome (in)stability and longevity assurance responses?

2.5 The omics connection between progeria and longevity

The aforementioned hypothesis raises the question whether different types of threats trigger a similar **survival response** or whether each stressor is counteracted by a tailored reaction. To gain further insight into the mechanisms that regulate mammalian longevity, we and others have previously quantified the parallels between the liver gene expression profiles of mutant dwarf and CR mice (that benefit from healthy aging and unusually long lifespan) and those of DNA repair-deficient progeroid animals (that age and die prematurely). Contrary to expectation, this comparative transcriptome analysis revealed significant, genome-wide expression associations between the progeroid and long-lived mice. Similar to previous studies, the most significantly overrepresented biological processes involved the suppression of the endocrine and energy pathways along with increased stress responses in both delayed and premature aging. Interestingly, a similar shift in gene expression towards somatic preservation as seen in NER progeria, mutant dwarfism, and CR also occurs with natural aging. For example, the majority of genes associated with immune, stress, and defense responses as well as programmed cell death were upregulated, whereas the majority of genes associated with growth, energy utilization, and lipid and carbohydrate metabolism were downregulated in the aged liver, kidneys, spleen, and lungs [26, 54]. Together, these findings have proved instrumental in identifying a common gene expression response to diverse stressors (Figure 2.2b). Furthermore, they challenged previous concepts for aging biomarkers; the great majority of genes with a central role in longevity assurance pathways did not represent biomarkers that would indicate biological age or longevity prospects. Instead, they appeared to be associated with responses to intrinsic or extrinsic stress factors.

Given that a preservative metabolic strategy triggered by CR or observed in long-lived dwarf mice has seemingly mainly beneficial effects on health and longevity, one cannot but wonder why this response is associated with NER progeroid animals or why it is not continually maximally switched on. The most plausible answers derive from the following considerations. Suppression of genes associated with the somatotrophic, lactotrophic, and thyrotrophic axes together with the upregulation of genes associated with stress and defense responses is indicative of a shift from growth to somatic maintenance [2]. This implies that in biological terms first priority after birth is growth and development to become adult and deliver progeny. Under favorable circumstances, the default strategy is maximal investment into growth and development, since such conditions permit reaching adulthood with damage levels that are easily tolerable. At this stage, overinvestment in protection and maintenance systems may be at the expense of resources available for growth and development and thus selected against [2]. Moreover, maximizing defenses under favorable circumstances may actively inhibit growth to (over)protect against cancer and may leave an organism unable to mount an even stronger defense when necessary. However, at life-threatening conditions such as food shortage but also stress, trauma, high fever, inflammation, or ischemia–reperfusion (latter three involving high levels of oxidative damage), it is unwise to spend scarce resources for growth. Furthermore, promoting growth and proliferation under adverse, damaging conditions would be dangerous in terms of the risk of cancer. The aforementioned emergency situations necessitate adjustments to survive the crisis by transiently prioritizing protection and maintenance at the temporary expense of growth. This is a key response strongly selected for in evolution with a high penalty rate, as it may be decisive for **survival of the fittest**, explaining its universal occurrence and early evolutionary origin. Therefore, this vital response is termed a **survival** response. In mammals, a shift from growth to somatic preservation may also function as a tumor suppressor mechanism and, accordingly, explain the diverse outcomes of distinct DNA repair deficiencies in human diseases. Together, these findings allowed us to propose that the progressive suppression of the insulin/IGF1 pathway and metabolism along with the upregulation of antioxidant and defense responses represents an adaptive response aiming at minimizing further damage by shifting the energy equilibrium from growth and proliferation to pathways that enhance somatic maintenance and promote organismal survival [12, 55] (Figure 2.2b). Beneficial effects of this response become apparent not only in long-lived pituitary mutant dwarf mice that show absence or greatly delayed occurrence of neoplasias and extended lifespan but also in NER-deficient animals where segmental progeria protects against cancer even in tissues that do not show age-related pathology. For instance, besides CR or mutant dwarf mice, progeroid Xpd^{TTD} mice, which have a milder TC-NER defect than $Csb^{m/m}/Xpa^{-/-}$, $Ercc1^{-/-}$ and $Ercc1^{-/\Delta-7}$ mice, are protected from tumorigenesis and, despite showing segmental progeria in a variety of tissues, show a number of paradoxically improved histopathologic changes related to a CR-like phenotype in other tissues such as lower incidence and/or severity of demyelination of the peripheral nerve, cataract, thyroid follicular distension, pituitary adenomas, and ulcerative dermatitis [56].

2.6 Triggering of systemic versus cell-autonomous features of the survival response

In mammals, the CR-, fasting-, or DNA damage-induced survival response involves systemic suppression of the somato-, lacto-, and thyrotrophic hormonal axes, which implicate the hypothalamus and pituitary as key endocrine organs. This raises the issue whether the survival response is determined by these organs or by every tissue in the body in a cell-autonomous manner. Recent studies have shown that UV light, which induces DNA lesions, such as CPDs, elicits dose-dependent CR-like expression changes in all types of cultured cells tested [57], indicating a cell-autonomous origin of the survival response. However, all DNA-damaging agents including UV induce also injury to other cellular biomolecules, complicating the identification of the type of damage responsible for triggering the observed hormonal changes. Fortunately, CPD lesions can be selectively removed by introduction of a DNA repair enzyme called photolyase [7]. Repair of this lesion simultaneously alleviated the response identifying CPD-type DNA lesions as the main culprit in the case of UV-induced damage [57]. This shows that DNA damage *per se* is able to act as one of the instigators of the survival mode. It is important to realize that CPD lesions serve as a prototype cytotoxic DNA injury that can persist in the genome. The response occurs in proliferating, quiescent, and postmitotic cells, indicating its universal, cell cycle independent nature. Thus, the ability of the growth-attenuating survival response to be triggered in cells throughout the body permits the occurrence of heterogeneity within organisms and even within individual organs and tissues. In this scenario, the hypothalamus/pituitary hormonal axis likely mediates the systemic component of the response controlled by the overall cellular status in the body.

How is the survival response triggered? At this stage, we are only left with speculation. The main stressors known to promote the survival response such as CR, fasting, and DNA damage interfere at multiple points with the basic cellular machinery: many aspects of energy metabolism, replication, and transcription, which all have multiple indirect effects on other vital processes. CPD lesions capable of triggering the survival mode represent persistent DNA lesions which are known to interfere with transcription elongation and require the aforementioned TCR system to allow resumption of the vital process of gene expression [57]. Mice with defects in this repair pathway exhibit multiple features of premature aging (like corresponding human patients), but at the same time, they also mount a systemic "survival" response, pointing to a connection between hampered transcription induced by DNA damage and activation of the survival mode. Thus, a state of hampered transcription elongation may initiate the signaling cascade that culminates in a cellular survival mode. Similarly, hampered cellular metabolism, for example, due to energy shortage caused by CR or fasting may slow down transcription and/or translation. Hampered replication as a trigger of the survival response may be somewhat less critical as the response also occurs in quiescent and postmitotic cells. Clearly, still much needs to be learned about the intracellular signaling systems that monitor key vital processes and in case of emergency switch on the survival program.

2.7 The omics connection between NER progeria, transcription, and longevity

Defects in NER and the ensuing genome instability have long been established as the underlying cause of mutations leading to skin cancer. However, and despite the tremendous progress that has been observed during the last few years, the links between NER defects and the developmental and metabolic abnormalities seen in NER disorders still remain obscure. For instance, it remains unknown how the accumulation of irreparable DNA lesions in NER-deficient animals leads to the suppression of key hormonal pathways, which molecules are the primary mediators of this response, or how perturbed DNA damage signaling leads to gene expression changes that involve dramatic changes in chromatin configuration. Besides DNA repair, earlier studies have shown that distinct NER factors play a role in transcription [22, 58] and upon stimulation, certain NER proteins are recruited to active promoters *in vitro* [59, 60]. These findings were crucial as they showed that NER factors may also function in nuclear processes (e.g., transcription) that are distinct from DNA repair. However, the *in vivo* relevance of NER-mediated transcription to the complex NER developmental disorders remained elusive, primarily due to difficulties in dissecting the dual role of NER in DNA repair and transcription in an intact organism. Fortunately, a series of functional genomics approaches were meant again to substantially advance our understanding on the functional role of NER in development and disease. Using an NER progeroid mouse model, we recently showed that key pathological features originate from defective transcription initiation of developmental stage- and tissue-specific gene expression programs *in vivo* [61, 62]. Surprisingly, those gene expression programs also involved longevity pathways that associate with the GH/IGF1 signaling and the thyrotroph and lactotroph pathways. For example, several NER factors, including ERCC1, were found to assemble together with the RNA POL II transcriptional machinery at the promoters of genes associated with growth *in vivo*. Thus, NER factors were shown to function beyond DNA repair in the transcriptional regulation of postnatal growth and differentiation. This adds a novel piece to the puzzle for understanding the causal basis of complex NER disorders. More importantly, however, it also suggests that NER factors may impinge on the transcriptional regulation of lifespan regulatory networks in ways other than those previously described (e.g., through the accumulation of irreparable DNA lesions).

2.8 Future perspectives

Recent years have witnessed exciting progress in understanding the molecular basis of aging and various components of the anticancer, antiaging lifespan-assurance survival pathways triggered by DR and other cues. We have now reached the stage at which translation of this accumulated knowledge into rational intervention strategies seems realistic. Two types of interventions (or better a combination of the two) can be envisaged. As aging is intimately associated with time-depended accumulation of damage in cellular biomolecules, for example, DNA, eroding the cellular machinery and functioning, one option is to reduce the induction or effects of the damaging molecular species. Obviously, it is advisable to avoid exogenous sources of cytotoxic and mutagenic types of damage. However, a likely continuous source of damage originates from ROS and their subsequent reaction products endogenously generated by mitochondrial respiration and lipid peroxidation. This source mostly coming from within may be modulated by adequate antioxidants and by altering metabolism. A second strategy may be artificially unlocking the potential of the survival program using ways more appealing than DR or less harmful than DNA damage. For instance, derivatives of resveratrol found in red wine and also other compounds may act as DR mimetics [63, 64]). Surprisingly, rapamycin, a potent immunosuppressant used by transplant patients has been recently shown to extend murine lifespan by 28–38% when fed late in life [65], presumably by inhibition of mTOR and/or S6L1 kinase signaling [66], processes with multiple links with the survival response. Although this agent has negative side effects, it is expected that the next era will witness significant advances in these directions.

References

1. Johnson FB, Sinclair DA, Guarente L. Molecular biology of aging. *Cell* 1999;96:291–302.
2. Kirkwood TB. Understanding the odd science of aging. *Cell* 2005;120:437–447.
3. Guarente L, Kenyon C. Genetic pathways that regulate ageing in model organisms. *Nature* 2000;408: 255–262.
4. Kenyon C. The plasticity of aging: Insights from long-lived mutants. *Cell* 2005;120:449–460.
5. Partridge L, Gems D. Mechanisms of ageing: Public or private? *Nat Rev Genet* 2002;3:165–175.
6. Mitchell JR, Hoeijmakers JH, Niedernhofer LJ. Divide and conquer: Nucleotide excision repair battles cancer and ageing. *Curr Opin Cell Biol* 2003;15:232–240.
7. Garinis GA, Mitchell JR, Moorhouse MJ, et al. Transcriptome analysis reveals cyclobutane pyrimidine dimers as a major source of UV-induced DNA breaks. *EMBO J* 2005;24:3952–3962.
8. Hoeijmakers JH. Genome maintenance mechanisms for preventing cancer. *Nature* 2001;411: 366–374.
9. Harper JW, Elledge SJ. The DNA damage response: Ten years after. *Mol Cell* 2007;28:739–745.

10. Hanawalt PC. Subpathways of nucleotide excision repair and their regulation. *Oncogene* 2002;21: 8949–8956.

11. Vijg J. Somatic mutations and aging: A re-evaluation. *Mutat Res* 2000;447:117–135.

12. Garinis GA, van der Horst GT, Vijg J, Hoeijmakers JH. DNA damage and ageing: New-age ideas for an age-old problem. *Nat Cell Biol* 2008;10:1241–1247.

13. Hasty P, Campisi J, Hoeijmakers J, van Steeg H, Vijg J. Aging and genome maintenance: Lessons from the mouse? *Science* 2003;299:1355–1359.

14. Garinis GA. Nucleotide excision repair deficiencies and the somatotropic axis in aging. *Hormones (Athens)* 2008;7:9–16.

15. Nance MA, Berry SA. Cockayne syndrome: Review of 140 cases. *Am J Med Genet* 1992;42:68–84.

16. Bootsma D, Kraemer KH, Cleaver JE, Hoeijmakers J. Nucleotide excision repair syndromes: Xeroderma pigmentosum, Cockayne syndrome and trichothiodystrophy. New York: McGraw-Hill Medical Publishing Division; 2002.

17. Vermeulen W, Rademakers S, Jaspers NG, et al. A temperature-sensitive disorder in basal transcription and DNA repair in humans. *Nat Genet* 2001;27:299–303.

18. Jaspers NG, Raams A, Silengo MC, et al. First reported patient with human ERCC1 deficiency has cerebro-oculo-facio-skeletal syndrome with a mild defect in nucleotide excision repair and severe developmental failure. *Am J Hum Genet* 2007;80:457–466.

19. Niedernhofer LJ, Garinis GA, Raams A, et al. A new progeroid syndrome reveals that genotoxic stress suppresses the somatotroph axis. *Nature* 2006;444:1038–1043.

20. Martin GM. Genetic modulation of senescent phenotypes in *homo sapiens*. *Cell* 2005;120:523–532.

21. van der Pluijm I, Garinis G, Brandt RMC, et al. Impaired genome maintenance suppresses the growth hormone–insulin-like growth factor 1 axis in mice with Cockayne syndrome. *PLoS Biol* 2006;5:23–38.

22. de Boer J, Andressoo JO, de Wit J, et al. Premature aging in mice deficient in DNA repair and transcription. *Science* 2002;296:1276–1279.

23. Murga M, Bunting S, Montana MF, et al. A mouse model of ATR-Seckel shows embryonic replicative stress and accelerated aging. *Nat Genet* 2009;41:891–898.

24. Hishiya A, Ito M, Aburatani H, Motoyama N, Ikeda K, Watanabe K. Ataxia telangiectasia mutated (atm) knockout mice as a model of osteopenia due to impaired bone formation. *Bone* 2005;37:497–503.

25. Marino G, Ugalde AP, Salvador-Montoliu N, et al. Premature aging in mice activates a systemic metabolic response involving autophagy induction. *Hum Mol Genet* 2008;17:2196–2211.

26. Schumacher B, Garinis GA, Hoeijmakers JH. Age to survive: DNA damage and aging. *Trends Genet* 2008;24:77–85.

27. van der Pluijm I, Garinis GA, Brandt RM, et al. Impaired genome maintenance suppresses the growth hormoneinsulin-like growth factor 1 axis in mice with Cockayne syndrome. *PLoS Biol* 2007;5:e2.

28. Longo VD, Finch CE. Evolutionary medicine: From dwarf model systems to healthy centenarians? *Science* 2003;299:1342–1346.

29. Barzilai N, Banerjee S, Hawkins M, Chen W, Rossetti L. Caloric restriction reverses hepatic insulin resistance in aging rats by decreasing visceral fat. *J Clin Invest* 1998;101:1353–1361.

30. Koubova J, Guarente L. How does calorie restriction work? *Genes Dev* 2003;17:313–321.

31. Spaulding CC, Walford RL, Effros RB. Calorie restriction inhibits the age-related dysregulation of the cytokines TNF-alpha and IL-6 in c3B10RF1 mice. *Mech Ageing Dev* 1997;93:87–94.

32. Yang H, Youm YH, Nakata C, Dixit VD. Chronic caloric restriction induces forestomach hypertrophy with enhanced ghrelin levels during aging. *Peptides* 2007;28:1931–1936.

33. Zhu M, Miura J, Lu LX, et al. Circulating adiponectin levels increase in rats on caloric restriction: The potential for insulin sensitization. *Exp Gerontol* 2004;39:1049–1059.

34. Heydari AR, Unnikrishnan A, Lucente LV, Richardson A. Caloric restriction and genomic stability. *Nucleic Acids Res* 2007;35:7485–7496.

35. Cho CG, Kim HJ, Chung SW, et al. Modulation of glutathione and thioredoxin systems by calorie restriction during the aging process. *Exp Gerontol* 2003;38:539–548.

36. Jia K, Levine B. Autophagy is required for dietary restriction-mediated life span extension in *C. elegans*. *Autophagy* 2007;3:597–599.

37. Grifantini K. Understanding pathways of calorie restriction: A way to prevent cancer? *J Natl Cancer Inst* 2008;100:619–621.

38. Alderman JM, Flurkey K, Brooks NL, et al. Neuroendocrine inhibition of glucose production and resistance to cancer in dwarf mice. *Exp Gerontol* 2009;44:26–33.

39. Colman RJ, Anderson RM, Johnson SC, et al. Caloric restriction delays disease onset and mortality in rhesus monkeys. *Science* 2009;325:201–204.

40. Li S, Crenshaw EB, 3rd, Rawson EJ, Simmons DM, Swanson LW, Rosenfeld MG. Dwarf locus mutants lacking three pituitary cell types result from mutations in the POU-domain gene Pit-1. *Nature* 1990;347:528–533.

41. Andersen B, Pearse RV, 2nd, Jenne K, et al. The Ames dwarf gene is required for Pit-1 gene activation. *Dev Biol* 1995;172:495–503.

42. Flurkey K, Papaconstantinou J, Miller RA, Harrison DE. Lifespan extension and delayed immune and collagen aging in mutant mice with defects in growth hormone production. *Proc Natl Acad Sci USA* 2001;98:6736–6741.

43. Zhou Y, Xu BC, Maheshwari HG, et al. A mammalian model for Laron syndrome produced by targeted disruption of the mouse growth hormone receptor/binding protein gene (the Laron mouse). *Proc Natl Acad Sci USA* 1997;94:13215–13220.

44. Holzenberger M, Dupont J, Ducos B, et al. Igf-1 receptor regulates lifespan and resistance to oxidative stress in mice. *Nature* 2003;421:182–187.

45. Kurosu H, Yamamoto M, Clark JD, et al. Suppression of aging in mice by the hormone Klotho. *Science* 2005;309:1829–1833.

46. Bartke A, Brown-Borg H. Life extension in the dwarf mouse. *Curr Top Dev Biol* 2004;63:189–225.

47. van de Ven M, Andressoo JO, Holcomb VB, et al. Adaptive stress response in segmental progeria resembles long-lived dwarfism and calorie restriction in mice. *PLoS Genet* 2006;2:e192.

48. Ramsey MM, Ingram RL, Cashion AB, et al. Growth hormone-deficient dwarf animals are resistant to dimethylbenzanthracine (DMBA)-induced mammary carcinogenesis. *Endocrinology* 2002;143:4139–4142.

49. Bartke A, Chandrashekar V, Bailey B, Zaczek D, Turyn D. Consequences of growth hormone (GH) overexpression and GH resistance. *Neuropeptides* 2002;36:201–208.

50. Groesbeck MD, Parlow AF, Daughaday WH. Stimulation of supranormal growth in prepubertal, adult plateaued, and hypophysectomized female rats by large doses of rat growth hormone: Physiological effects and adverse consequences. *Endocrinology* 1987;120:1963–1975.

51. Brown-Borg HM, Rakoczy SG. Catalase expression in delayed and premature aging mouse models. *Exp Gerontol* 2000;35:199–212.

52. Brown-Borg HM, Rakoczy SG. Growth hormone administration to long-living dwarf mice alters multiple components of the antioxidative defense system. *Mech Ageing Dev* 2003;124:1013–1024.

53. Cao SX, Dhahbi JM, Mote PL, Spindler SR. Genomic profiling of short- and long-term caloric restriction effects in the liver of aging mice. *Proc Natl Acad Sci USA* 2001;98:10630–10635.

54. Schumacher B, van der Pluijm I, Moorhouse MJ, et al. Delayed and accelerated aging share common longevity assurance mechanisms. *PLoS Genet* 2008;4:e1000161.

55. Garinis GA, Schumacher B. Transcription-blocking DNA damage in aging and longevity. *Cell Cycle* 2009;8:2134–2135.

56. Wijnhoven SW, Beems RB, Roodbergen M, et al. Accelerated aging pathology in ad libitum fed Xpd(TTD) mice is accompanied by features suggestive of caloric restriction. *DNA Repair (Amst)* 2005;4:1314–1324.

57. Garinis GA, Uittenboogaard LM, Stachelscheid H, et al. Persistent transcription-blocking DNA lesions trigger somatic growth attenuation associated with longevity. *Nat Cell Biol* 2009;11:604–615.

58. Citterio E, Van Den Boom V, Schnitzler G,. ATP-dependent chromatin remodeling by the Cockayne syndrome B DNA repair-transcription-coupling factor. *Mol Cell Biol* 2000;20:7643–7653.

59. Le May N, Egly JM, Coin F. True lies: The double life of the nucleotide excision repair factors in transcription and DNA repair. *J Nucleic Acids* 2010;2010:Article ID 616342:10 pages.

60. Le May N, Mota-Fernandes D, Velez-Cruz R, Iltis I, Biard D, Egly JM. NER factors are recruited to active promoters and facilitate chromatin modification for transcription in the absence of exogenous genotoxic attack. *Mol Cell* 2010;38:54–66.

61. Kamileri I, Karakasilioti I, Garinis GA. Nucleotide excision repair: New tricks with old bricks. *Trends Genet* 2012;28:566–573.

62. Kamileri I, Karakasilioti I, Sideri A, et al. Defective transcription initiation causes postnatal growth failure in a mouse model of nucleotide excision repair (NER) progeria. *Proc Natl Acad Sci USA* 2012;109:2995–3000.

63. Baur JA, Sinclair DA. Therapeutic potential of resveratrol: The in vivo evidence. *Nat Rev Drug Discov* 2006;5:493–506.

64. Ingram DK, Zhu M, Mamczarz J, et al. Calorie restriction mimetics: An emerging research field. *Aging Cell* 2006;5:97–108.

65. Harrison DE, Strong R, Sharp ZD, et al. Rapamycin fed late in life extends lifespan in genetically heterogeneous mice. *Nature* 2009;460:392–395.

66. Selman C, Tullet JM, Wieser D, et al. Ribosomal protein S6 kinase 1 signaling regulates mammalian life span. *Science* 2009;326:140–144.

3 Protein Quality Control Coming of Age

Silke Meiners

Comprehensive Pneumology Center (CPC), University Hospital, Ludwig-Maximilians-University, Helmholtz Zentrum München, Member of the German Center for Lung Research (DZL), Munich, Germany

3.1 Introduction

At the molecular level, aging is characterized by the accumulation of damaged DNA, proteins, lipids, and carbohydrates in cells and tissues. Molecular damage accumulates as the number of noxious stressors increases over time, while repair and stress response systems become progressively impaired. Aging may thus be regarded as the outcome of the balance between damage and repair as recently reviewed by Haigis and Yankner and as outlined in Figure 3.1 [1]. Accordingly, counterbalancing stress and improving cellular repair systems is emerging as a potential strategy to ameliorate age-related disorders [2]. While the cellular response to accumulating DNA damage and telomere dysfunction is discussed in Chapters 2 and 4, this chapter focuses on the cellular stress response related to protein impairment, that is, protein quality control, and its age-related alterations.

Maintenance of the functional homeostasis of all proteins within the cell is called proteostasis. Its proper function is essential for the survival of cells and organisms [3]. A central part of proteostasis is protein quality control. It involves the tight supervision of protein synthesis, correct protein maturation and folding, as well as the timely disposal of unwanted and damaged proteins. Chaperones foster correct three-dimensional folding of proteins, while unwanted and misfolded proteins are disposed via the ubiquitin–proteasome and lysosomal autophagy pathways. Challenges to protein quality control are rapidly sensed on the level of chaperone availability in all cellular compartments and trigger a specific and conserved cellular stress response that allows the cell either to adapt to altered protein homeostasis or to undergo controlled cell death, namely, apoptosis. The unfolded protein response (UPR) of the endoplasmic reticulum (ER) is a particularly well-characterized example for the induction of a compartment-specific stress response.

Aging is a particular challenge for the proteostasis network of the cell: Not only that misfolded and impaired proteins are accumulating over time, but also protein quality control pathways become functionally impaired in the process of aging [4]. It has been proposed that "a collapse of proteostasis represents an early molecular event in the aging process

Molecular Aspects of Aging: Understanding Lung Aging, First Edition. Edited by Mauricio Rojas, Silke Meiners and Claude Jourdan Le Saux.

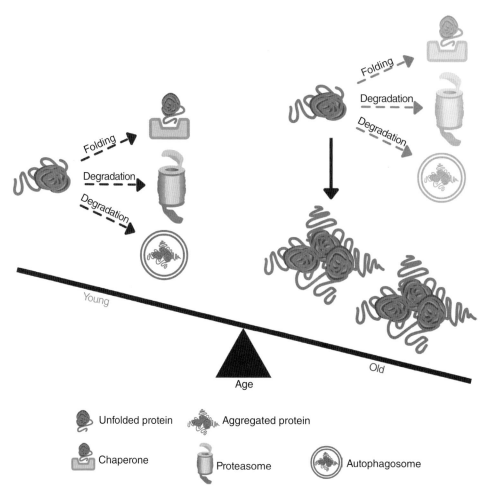

Figure 3.1 Imbalanced proteostasis upon aging. While the proteostasis network is well balanced in young cells due to proper chaperone-mediated protein refolding or protein degradation via the autophagosome or proteasome pathways, this balance is tipped in favor of protein damage and proteotoxic protein aggregation in aging cells due to impaired chaperone networks and protein degradation pathways.

that amplifies protein damage in age-associated diseases of protein conformation" [5]. Accumulating protein damage has been mainly ascribed to an increased load of reactive oxygen species (ROS) over time and is observed in several age-related diseases. ROS react either with the peptide bonds of the protein backbone resulting in peptide truncation or mediate oxidation, carbonylation, or direct glycation of amino acid side chains, and thereby contribute to protein misfolding and inactivation [6]. Age-related changes of protein quality control have been observed on all levels as discussed in the following. Accordingly, aging is a particular risk factor for sporadic protein quality control diseases such as neurodegenerative and cardiovascular disorders and cataract formation [4]. Moreover, premature aging of tissues as, for example, shown for some chronic lung diseases and HIV infection is an emerging new theme in clinical research. Understanding the pathways that initiate premature aging might help to identify new targets for therapeutic disease control.

3.2   The aging molecular chaperone network

Adaptation of the chaperone network during aging is crucial for proper cellular function as the number of proteotoxic insults increases over time and contributes to the accumulation of damaged and misfolded proteins [4]. By definition, chaperones are proteins that transiently help other proteins to acquire their final functional conformation [7]. Synthesis of functional proteins involves not only the translation of mRNA into peptide chains by the ribosome factory but also the fine-tuned folding of proteins into a complex three-dimensional conformation which then enables proper protein function. This is achieved mainly cotranslationally and with the help of distinct sets of chaperones [8]. Depending on the cellular localization of the protein (e.g., nucleus, mitochondria, plasma membrane, ER), chaperones differ, ranging from ER-resident to mitochondrial-specific chaperone families [7, 9]. Chaperones also assist in the regulation of protein localization and protein–protein interactions, thereby contributing to functional protein networks [3]. Three-dimensional folding of mature proteins is endangered by stress-induced protein modifications, lack of oligomeric assembly partners, or denaturation and results in protein misfolding [4]. These proteins expose then hydrophobic surfaces that complex with normal proteins to form protein aggregates, thereby challenging cellular function, a process called proteotoxicity. Molecular chaperones prevent protein aggregation by binding to these hydrophobic residues, thereby facilitating correct folding in an ATP-dependent manner [8]. Ensuring proper protein folding by molecular chaperone networks thus represents the first level of protein quality control.

As chaperones have been originally described to be induced by heat shock, they have been termed heat shock proteins (Hsps) and are grouped according to their molecular weight. There are five major and well-conserved families of chaperones, namely, Hsp100s, Hsp90s, Hsp70s, Hsp60, and small Hsps (sHsps) [10]. Molecular chaperones are promiscuous in their binding to unfolded proteins and have a high affinity for hydrophobic protein patches. sHsps only bind nonnative proteins, thereby holding this unfolded conformation. In contrast, the other Hsps not only bind but also catalyze refolding of the partly folded protein into its native state in an ATP-dependent manner. While the holdases are often only expressed at conditions of stress, the so-called foldases come as stress-induced and constitutive versions [10].

Expression of Hsps is rapidly induced via the specific transcription factor heat shock factor-1 (Hsf-1) which transcriptionally activates heat shock-responsive genes via binding to a conserved heat shock element (HSE) in their promotor regions [11]. The activation of Hsf-1 needs to be rapid and involves its trimerization. According to the chaperone displacement model, Hsf-1 is kept in an inactive conformation by complexing with chaperones, most prominently Hsp90. In cells subjected to heat shock or other proteotoxic conditions, increased protein misfolding liberates Hsp90 from Hsf-1, thereby facilitating Hsf-1 trimerization and activation. In addition, Hsf-1 seems to be regulated by a thermosensitive noncoding RNA sensor and by oxidative modification of key cysteine residues which might account for the extremely rapid activation of Hsf-1 within few minutes [12]. Stress-activated Hsf-1 can be further regulated by various posttranslational modifications such as phosphorylation, sumoylation, and acetylation to fine-tune and precisely regulate the levels of chaperones in the cell [10]. Hsf-1 can thus be regarded as a master regulator of proteotoxic stress [11].

Expression of Hsf-1 and proper induction of Hsp responses have been shown not only to counteract protein misfolding and aggregation but also to affect life span in the model organism *C. elegans* [13, 14]. Loss of Hsf-1 function decreased life span, while the induction of Hsf-1 counteracted the age-dependent collapse of the proteostasis network induced in thermosensitive mutants [5]. Of note, the detrimental effects of reduced Hsf-1 expression on longevity were more pronounced than that of individual molecular chaperones, suggesting an important role for chaperone networks rather than for single Hsps in aging. It is also worth mentioning that the two central pathways for controlling cell growth and longevity, namely, the insulin growth factor and the mammalian target of rapamycin (mTOR) signaling pathways, directly regulate Hsf-1 activity [15, 16]. This link between the aging pacemaker pathway mTOR [17] and the chaperone network via Hsf-1 suggests a coordinated regulation of chaperone availability in response to the nutrient state of the aging cell.

3.3 Protein degradation pathways in aging

In addition to the accumulation of misfolded and dysfunctional proteins over time, the two main protein degradation systems, the ubiquitin–proteasome and the lysosomal autophagy pathways, become functionally impaired in the process of aging [4]. These two disposal systems not only perform housekeeping tasks of normal protein turnover but also quality control tasks that ensure rapid disposal of dysfunctional proteins. The essential function of the ubiquitin–proteasome pathway for controlled regulation of protein turnover in signaling and maintaining cell function has been publicly acknowledged in 2004, when Aaron Ciechanover, Avram Hershko, and Irvine Rose were awarded the Nobel Prize for their first description of the ubiquitin–proteasome system.

3.3.1 Lysosomal autophagy pathway

Consistent with the notion of protein quality control compartments coming of age, reduced function of the lysosomal autophagy pathway has been observed in aged tissues and age-related disorders [18, 19].

This degradation pathway comprises delivery of cytoplasmic cargo to the lysosome via autophagic vesicles and subsequent bulk degradation of the cargo in the lysosome. Three different forms of autophagy have been identified: chaperone-mediated autophagy (CMA), allowing selective and chaperone-assisted lysosomal degradation of specific proteins; microautophagy, involving direct sequestration of cytosolic components by invagination of the lysosomal membrane; and macroautophagy, an unspecific degradation process of long-lived or aggregated proteins and whole organelles often referred to simply as autophagy [20]. In autophagy, the cargo is sequestered inside a double-membrane vesicle. An autophagosomal isolation membrane is initiated by vesicle nucleation from the ER, the outer mitochondrial membrane, or the plasma membrane. Vesicles expand and form a phagophore. The autophagosome is formed upon fusion of the edges of the phagophore. Fusion with the lysosome results in the formation of autolysosomes where the captured material is degraded together with the inner vesicle membrane by acid hydrolases [19]. Degradation products such as free amino and fatty acids can be recycled for macromolecule synthesis or used for ATP synthesis via the tricarboxylic acid cycle, contributing to the essential prosurvival function of autophagy [20]. This critical role of autophagy in the mobilization of intracellular energy resources makes it a prime target for regulation by growth factor signaling pathways which are integrated on the level of mTOR kinase activation [21]. mTOR signaling is the major inhibitory signal that shuts down autophagy in the presence of growth factors and abundant nutrients. *Vice versa*, impaired growth factor signaling and nutrient deprivation as well as multiple forms of cellular stress such as nutrient and energy deprivation, hypoxia, redox stress, protein aggregates, damaged organelles, and intracellular pathogens induce autophagy catabolism [22]. This strongly indicates that autophagy is a major protective mechanism of the cell which helps to eliminate damaged and harmful components including organelles, viruses, and aggregated proteins [19].

On the molecular level, more than 30 genes – the so-called autophagy-related (ATG) genes – have been identified to be essential for the execution of all autophagy steps [20]. These genes are intricately regulated on the expressional and posttranslational level by growth factor and stress-related stimuli [22]. Reduced expression of ATG genes has been observed in aged tissues and age-related disorders as only recently comprehensively reviewed [18, 19]. Experimental impairment of autophagy in model organisms reduces life span and induces premature aging in worms, flies, and mice [18, 19]. This does not come as a surprise as not only the catabolic recycling function of autophagy is lost but also its essential role in protein quality control. The twofold role of autophagy is well manifested by the finding that the systemic knockout of ATG genes in mice is embryonically lethal – probably due to the nutrient recycling defect. Tissue-specific knockouts of ATG genes, however, reflect hallmarks of protein quality diseases such as accumulation of modified and aggregated proteins [18]. *Vice versa*, enhancing the autophagic flux in the cell, as, for example, by caloric restriction (which represents the most physiological inducer of autophagy), extends life span and counteracts age-related disease onset [19, 23].

A selective form of autophagy involved in the specific degradation of soluble cytosolic proteins is CMA. Here, the cytosolic cargo is identified via binding of the constitutive Hsp70 protein (Hsc70) to a specific interaction motif of the substrate [24]. Subsequently, the substrate–Hsc70 complex interacts with the cytoplasmic tail of the lysosome-membrane protein type 2A (LAMP-2A) and is translocated into the lysosome for degradation [25]. As about 30% of cytosolic proteins contain the consensus motif for CMA targeting, it can be envisioned that this specific autophagosomal degradation pathway of single proteins contributes significantly to cellular protein quality control [26]. Indeed, dysfunction of CMA has been observed in several diseases and contributes to altered protein homeostasis in nephropathies, neurodegenerative disorders, and lysosomal storage diseases [27]. Recent research also suggests that CMA might be involved in aging processes as indicated by a decreased LAMP-2A content in aged cells [28].

While altered autophagy function has been primarily observed in age-related disorders of nondividing tissue such as the brain and the heart, increased autophagy was recently observed in a highly proliferative organ, the lung. Patients with chronic obstructive pulmonary disease (COPD) show prominent signs of enhanced autophagy [29]. Upregulation of autophagy might be associated with the pathological remodelling of the organ as previously suggested for the failing heart [19, 30]. It is still a matter of debate whether such induction of autophagy is also related to premature aging as recently observed in COPD patients and discussed in Chapter 13. One drawback of most studies on autophagy is, however, that there is no discrimination of the different forms of autophagy. It is well conceivable that the accelerated autophagosomal degradation of mitochondria – known as mitophagy – has a different impact on cellular function than increased bulk degradation of cytosolic proteins. Further research will be required to dissect different autophagy pathways, their respective cargo, and their particular regulation with regard to age- and disease-related disorders.

3.3.2 Ubiquitin–proteasome system

Proteasome function has been reported to be altered in the process of aging on multiple levels such as expression, assembly, subunit composition, and posttranslational modifications (see section 3.3.2.1). The essential role of the ubiquitin–proteasome system for proteostasis in general and aging processes in particular is reflected by the fact that more than 90% of all cellular proteins are specifically degraded by the proteasome [31]. Proper proteasome function is thus essential for numerous cellular processes such as protein turnover and quality control, cell growth and signaling, and immune response and antigen presentation [32, 33, 34]. According to its functional relevance, the proteasome constitutes about 1% of total cellular protein content [31]. Proteasomal protein degradation takes place in the nucleus and cytoplasm [35]. The degradation of cellular proteins by the ubiquitin–proteasome system needs to be highly selective. For that, degradation-prone proteins are first tagged with ubiquitin chains and subsequently degraded by the proteasome. Ubiquitination proceeds along a cascade of enzymatic reactions in which ubiquitin is first activated by the ubiquitin-activating enzyme E1, then conjugated with the help of an E2 ubiquitin-conjugating enzyme, and finally covalently linked to the substrate by a specific E3 ubiquitin ligase. Subsequently, further ubiquitin moieties are transferred to a particular lysine residue of the previously conjugated ubiquitin molecule to form a poly-ubiquitin chain [36]. There are only few E1 enzymes, several E2s, and multiple classes of E3s. The 26S proteasome is able to bind and degrade these polyubiquitinated proteins. It is composed of one barrel-shaped 20S catalytic core and one or two cap-like 19S regulatory complexes. While the regulators facilitate binding and deubiquitination, as well as ATP-dependent unfolding of substrates, the core particle hydrolyzes proteins with the help of three proteolytic active sites into 3–25 amino acid-long peptides [37]. These peptides are either further hydrolyzed into amino acids by secondary hydrolases for recycling of amino acids or used for MHC class I antigen presentation to define the cell's self towards the immune system [38].

Regulation of proteasomal proteolysis is less well studied than autophagosomal protein degradation. Given the central role of this system for cellular protein turnover and the coordinated and complex assembly of the proteasomal degradation machinery from more than 30 proteins [39], it is easy to imagine that the regulation of the proteasomal degradation complex is neither rapid nor excessive. The enzymatic activity of the proteasome is mainly regulated by association of the 20S catalytic core complex with cap-like regulators such as the 19S complex, the proteasome activators 11 (also called PA28) and 200 (PA200), which seem to have either cell type or compartment-specific functions [40]. In yeast, the regulation of proteasomal gene expression and assembly is governed by a single master transcription factor, RPN4 [41]. In mammalians, however, a single transcriptional master regulator has not been identified so far. The two stress-responsive transcription factors Nrf1 and Nrf2 have been reported to induce transcriptional expression of single proteasomal subunits linking proteasome expression to oxidative and proteotoxic stresses [42, 43]. Specific regulation of substrate turnover takes place on the level of ubiquitin-conjugating E3 ligases or – as a newly emerging theme – on the level of specific deubiquitinating enzymes, so-called DUBs [44]. Specificity of E3 ligases is not restricted to a single substrate but promiscuous as one ligase usually has several substrates. *Vice versa*, the degradation of one substrate is regulated by several E3 complexes depending on the cellular context [36]. Misfolded and translational junk proteins are also degraded via the ubiquitin–proteasome system [45]. As outlined previously, proteins are targeted for proteasomal degradation in case chaperones are unable to successfully restore functional protein folding. This concept is known as the protein triage model for quality control [46]. One of the major enzymes for ubiquitination of cytosolic misfolded proteins is the Hsp70-dependent E3-ligase carboxy terminus of Hsp70-interacting protein (CHIP) which binds to Hsp-associated misfolded substrates [47]. CHIP has been shown to be relevant for degradation of misfolded proteins (such as tau) in neurodegenerative as well as in cardiovascular diseases [48, 49]. There is, however, more than one so-called quality control E3 ligase: Several of such ligases are associated with the ER to facilitate ubiquitin-mediated degradation of ER-associated misfolded proteins, some have been implicated in mitochondrial protein quality control, while others take over quality control jobs in the cytosol or nucleus [50]. The essential role of the proteasome for protein quality control in general and degradation of oxidatively and misfolded proteins in particular is well established [51, 52].

3.3.2.1 The aging proteasome system

Several studies described reduced expression of proteasomal genes in aged cells and tissues of flies, rats, and men [53, 54, 55]. *Vice versa*, conserved proteasome expression was observed in cells of long-lived centenarians [56]. Accordingly, caloric restriction – the only real anti-age measure – counteracts loss of proteasome function and contributes to elevated levels of proteasome subunits and regulators [53]. In addition to the regulation of proteasomal gene expression, impairment of proteasome activity has been observed on several levels and contributes to reduced longevity: in worms, loss of a regulator of proteasome activity shortened life span [57]. In flies, disturbed assembly of 26S proteasomes was observed upon aging and genetically forced impairment of this assembly reduced life span [58, 59]. In addition, oxidative and chemical modification of catalytic and noncatalytic proteasomal subunits has been reported to directly impair proteasomal activity at conditions of oxidative stress and aging [51, 60, 61]. Moreover, it has been proposed that cross-linked protein aggregates, as observed in aged tissue and cells, inhibit proteasome activity [62]. *In vitro*, the inhibition of proteasome activity increased the senescence of fibroblasts, while the activation of proteasome activity delayed the senescence program [63, 64]. A causal role for functional proteasome activity as an essential factor for longevity is also indicated by genetic studies in *C. elegans* and Drosophila when components of the 19S proteasome regulator were either knocked out or overexpressed and induced accelerated organismal aging or conferred longevity, respectively [58, 65].

In contrast to impaired proteasome function in aging, no general age-related impairment of the enzymatic ubiquitination machinery has been observed so far. Single E3 ligases, however, have been implicated in the regulation of cellular senescence and organismal aging. One prominent example is the E3 ligase WWW-1, which is a positive regulator of life span in *C. elegans* in response to caloric restriction [66]. On the other hand, the E3 ligase Smurf2 seems to be a negative regulator as upregulation of Smurf2 activates and loss of Smurf2 impairs the senescence program of the cell [67, 68]. Knockout of the protein quality ligase CHIP in mice accelerates aging and is accompanied by impaired proteostasis [48]. This observation clearly indicates the tight connection between protein quality control and aging.

Only recently, knockdown of the deubiquitinase PSMD14 has been shown to induce cell cycle arrest and senescence, suggesting that age-related regulation of DUBs emerges as a new paradigm for the control of ubiquitin-mediated protein degradation in aging [69].

3.4 Compartment-specific protein quality control

Dysfunctional proteins can arise in all cellular compartments such as the plasma membrane, cytosol, nucleus, endoplasmic reticulum (ER), and mitochondria in aging. Nuclear proteins can be directly degraded by nuclear proteasomes or as part of nucleophagy [70]. Due to their ancestral bacterial nature, mitochondria contain their own protein quality control system consisting of mitochondrial chaperones and some ancestral proteolytic systems [9]. Bulk degradation of dysfunctional mitochondria takes place via cooperative action of the lysosomal and proteasomal pathways in a process called mitophagy [71].

As nearly one third of all proteins are secreted via the ER, this organelle is equipped with a highly specialized protein quality control involving special ER-resident chaperones. Misfolded proteins are rapidly recognized by ER-associated ubiquitin ligases and are ejected into the cytosol for ubiquitin-mediated proteasomal degradation. This process is termed ER-associated degradation pathway (ERAD) [72]. It is complemented by autophagy-mediated bulk disposal of parts of the dysfunctional ER [73]. In addition, misfolded proteins trigger the ER-specific UPR, a coordinated stress response program that adjusts the capacity of folding and disposal of the ER [74, 75]. Sequestering of chaperones, for example, BIP, by misfolded proteins in the ER is rapidly sensed by three ER stress sensors – the inositol-requiring transmembrane kinase/endoribonuclease 1 (IRE1), the double-stranded RNA (PKR)-activated protein kinase-like eukaryotic initiation factor 2a kinase (PERK), and the activating transcription factor-6 (ATF6) – and launches the three-armed adaptive stress response of the ER. While the activation of PERK attenuates protein translation thereby decreasing protein reloading of the ER, IRE1 and ATF6 concertedly activate UPR-specific transcription to increase the ER folding capacity and function. Activation of this complex stress response aims to restore protein homeostasis of the ER [76]. When they fail, however, for example, upon severe or chronic ER stress, apoptosis is switched on [77]. For the details of the ER-specific protein quality control the reader is referred to the excellent reviews cited above.

3.4.1 The aging ER stress response

Impaired ER stress response has been associated with a variety of diseases such as neurodegenerative, cardiovascular, and metabolic disorders, cancer, and inflammatory diseases by means of mouse models and genetic association studies [76]. The causal contribution of the UPR for disease onset and progression has also been clearly shown for several hereditary conformational diseases. The primary cause of diseases such as cystic fibrosis and alpha-1 antitrypsin (AAT) deficiency is the expression of mutant alleles of the cystic fibrosis transmembrane conductance regulator (CFTR) and AAT, respectively, which activate the UPR, thereby leading to a variety of pathophysiological consequences [78]. The role of the ER stress response in aging has been mainly implicated in age-related neurodegenerative diseases and diabetes. Components of the adaptive UPR, such as the ER-resident chaperones BIP and calnexin, and the signaling component PERK functionally decline with age [79]. This is accompanied by structural changes of the ER and disordering of the well-structured ER cisternae in aged cells. *Vice versa*, components of the apoptotic arm of the UPR, for example, CHOP, are found to be elevated in aged rats, suggesting a shift in the balance of protective versus apoptotic signals in favor of apoptotic UPR responses during aging [79]. According to that, aged mice are more sensitive to virus-induced lung fibrosis possibly due to higher levels of UPR-mediated apoptosis in lung epithelial cells [80]. Increased ER stress resistance, however, is observed in long-lived mouse and worm mutants [81, 82]. The bidirectional interplay of the mTOR and the UPR pathways also strongly indicate a coordinate regulation of mTOR growth factor signaling and protein quality control of the ER in the aging cell [83].

3.5 Conclusion

Protein quality control maintains proteome integrity and defends the cell from proteotoxic damage. Importantly, the different arms of the protein quality control network are tightly interconnected and may partly compensate for each other as exemplified by the interplay of chaperones, proteasome, and autophagy pathways [84] and by the concerted regulation via the mTOR growth factor signaling pathway. Protein damage accumulates more easily in postmitotic cells that are unable to dilute misfolded and aggregated proteins by cell division. These aggregates may then contribute to the classical age-dependent disorders of the brain, eyes, and heart [85]. Whether accumulating protein damage also contributes to premature aging of rapidly dividing and highly regenerative tissues, such as the lungs, is still unknown.

Impairment of protein quality control may take place in various tissues and cellular compartments including stem and immune cells. Age-related impairment of protein quality control pathways in stem cells has been proposed to add to overall organ and organismal aging [86]. Harnessing the proteostasis network in immune cells, as shown for proteasome function in T cells [87], might add to the detrimental and systemic effects of aging [88]. Of note, the immune aspect of the aging proteostasis network has largely been neglected so far: As proteasome function is essential for MHC class I antigen generation [34], altered proteasomal activity might affect the MHC class I antigen repertoire that is displayed on the surface of an aged cell. This adds to the risk of autoimmunity, a phenomenon which has also been shown to increase with aging (see Chapter 7). In addition, the autophagy pathway contributes to immune surveillance of pathogens, to MHC class II presentation of cytoplasmic and nuclear antigens, and possibly to cross-presentation pathways [89]. Altered activity of the lysosomal autophagy pathway may thus add to an altered immune surveillance of the immune system in aging as outlined in Chapter 7.

As summarized in Figure 3.1, the proteostasis network is well balanced in young cells, while this balance is tipped in favor of protein damage and proteotoxic protein aggregation in aging cells due to impaired chaperone networks and protein degradation pathways. Given the myriad of damaging insults and the complexity of protein control mechanisms, the tightly balanced protein repair system of the cell can be affected on multiple levels according to the pleiotropic nature of aging. While it seems not feasible to diminish the number of damaging insults while we age, the age-dependent impairment of the proteostasis network might be counteracted as recently suggested for protein conformational diseases [90]. Enhancing protein quality control systems may thus serve as a potential strategy to ameliorate age-related disorders.

References

1. Haigis, MC, Yankner, BA The aging stress response. *Molecular cell* 2010;40:333–44.
2. Kourtis, N, Tavernarakis, N Cellular stress response pathways and ageing: intricate molecular relationships. *The EMBO journal* 2011;30:2520–31.

3. Balch, WE, Morimoto, RI, Dillin, A, Kelly, JW Adapting proteostasis for disease intervention. *Science* 2008;319:916–9.
4. Morimoto, RI Proteotoxic stress and inducible chaperone networks in neurodegenerative disease and aging. *Genes & development* 2008;22:1427–38.
5. Ben-Zvi, A, Miller, EA, Morimoto, RI Collapse of proteostasis represents an early molecular event in *Caenorhabditis elegans* aging. *Proceedings of the national academy of sciences of the united states of America* 2009;106:14914–9.
6. Berlett, BS, Stadtman, ER Protein oxidation in aging, disease, and oxidative stress. *The journal of biological chemistry* 1997;272:20313–6.
7. Hartl, FU, Hayer-Hartl, M Converging concepts of protein folding in vitro and in vivo. *Nature structural & molecular biology* 2009;16:574–81.
8. Hartl, FU, Bracher, A, Hayer-Hartl, M Molecular chaperones in protein folding and proteostasis. *Nature* 2011;475:324–32.
9. Voos, W Chaperone-protease networks in mitochondrial protein homeostasis. *Biochimica et biophysica acta* 2013;1833:388–99.
10. Richter, K, Haslbeck, M, Buchner, J The heat shock response: life on the verge of death. *Molecular cell* 2010;40:253–66.
11. Akerfelt, M, Morimoto, RI, Sistonen, L Heat shock factors: integrators of cell stress, development and lifespan. *Nature reviews molecular cell biology* 2010;11:545–55.
12. Anckar, J, Sistonen, L Regulation of HSF1 function in the heat stress response: implications in aging and disease. *Annual review of biochemistry* 2011;80:1089–115.
13. Hsu, A-L, Murphy, CT, Kenyon, C Regulation of aging and age-related disease by DAF-16 and heat-shock factor. *Science* 2003;300:1142–5.
14. Morley, JF, Morimoto, RI Regulation of longevity in *Caenorhabditis elegans* by heat shock factor and molecular chaperones. *Molecular biology of the cell* 2004;15:657–64.
15. Chou, S-D, Prince, T, Gong, J, Calderwood, SK mTOR is essential for the proteotoxic stress response, HSF1 activation and heat shock protein synthesis. *PloS one* 2012;7:e39679.
16. Chiang, W-C, Ching, T-T, Lee, HC, Mousigian, C, Hsu, A-L HSF-1 regulators DDL-1/2 link insulin-like signaling to heat-shock responses and modulation of longevity. *Cell* 2012;148:322–34.
17. Kapahi, P, Chen, D, Rogers, AN, et al. With TOR, less is more: a key role for the conserved nutrient-sensing TOR pathway in aging. *Cell metabolism* 2010;11:453–65.
18. Rubinsztein, DC, Mariño, G, Kroemer, G Autophagy and aging. *Cell* 2011;146:682–95.
19. Levine, B, Kroemer, G Autophagy in the pathogenesis of disease. *Cell* 2008;132:27–42.
20. He, C, Klionsky, DJ Regulation mechanisms and signaling pathways of autophagy. *Annual review of genetics* 2009;43:67–93.
21. Laplante, M, Sabatini, DM mTOR signaling in growth control and disease. *Cell* 2012;149:274–93.
22. Kroemer, G, Mariño, G, Levine, B Autophagy and the integrated stress response. *Molecular cell* 2010;40:280–93.
23. Colman, RJ, Anderson, RM, Johnson, SC, et al. Caloric restriction delays disease onset and mortality in rhesus monkeys. *Science* 2009;325:201–4.
24. Terlecky, SR, Chiang, HL, Olson, TS, Dice, JF Protein and peptide binding and stimulation of in vitro lysosomal proteolysis by the 73-kDa heat shock cognate protein. *The journal of biological chemistry* 1992;267:9202–9.
25. Kaushik, S, Cuervo, AM Chaperone-mediated autophagy: a unique way to enter the lysosome world. *Trends in cell biology* 2012;22:407–17.
26. Arias, E, Cuervo, AM Chaperone-mediated autophagy in protein quality control. *Current opinion in cell biology* 2011;23:184–9.
27. Kon, M, Cuervo, AM Chaperone-mediated autophagy in health and disease. *FEBS letters* 2010;584:1399–404.
28. Cuervo, AM, Dice, JF Age-related decline in chaperone-mediated autophagy. *The journal of biological chemistry* 2000;275:31505–13.
29. Chen, Z-H, Kim, HP, Sciurba, FC, et al. Egr-1 regulates autophagy in cigarette smoke-induced chronic obstructive pulmonary disease. *PloS one* 2008;3:e3316.
30. Ryter, SW, Chen, Z, Kim, HP, Choi, AMK Autophagy in chronic obstructive pulmonary disease. *Autophagy* 2009;5:235–7.
31. Rock, KL, Gramm, C, Rothstein, L, et al. Inhibitors of the proteasome block the degradation of most cell proteins and the generation of peptides presented on MHC class I molecules. *Cell* 1994;78:761–71.

32. Geng, F, Wenzel, S, Tansey, WP Ubiquitin and proteasomes in transcription. *Annual review of biochemistry* 2012;81:177–201.

33. Goldberg, AL Protein degradation and protection against misfolded or damaged proteins. *Nature* 2003;426:895–9.

34. Kloetzel, P-M, Ossendorp, F Proteasome and peptidase function in MHC-class-I-mediated antigen presentation. *Current opinion in immunology* 2004;16:76–81.

35. Wójcik, C, DeMartino, GN Intracellular localization of proteasomes. *The international journal of biochemistry & cell biology* 2003;35:579–89.

36. Kerscher, O, Felberbaum, R, Hochstrasser, M Modification of proteins by ubiquitin and ubiquitin-like proteins. *Annual review of cell and developmental biology* 2006;22:159–80.

37. Finley, D Recognition and processing of ubiquitin-protein conjugates by the proteasome. *Annual review of biochemistry* 2009;78:477–513.

38. Yewdell, JW, Reits, E, Neefjes, J Making sense of mass destruction: quantitating MHC class I antigen presentation. *Nature reviews immunology* 2003;3:952–61.

39. Murata, S, Yashiroda, H, Tanaka, K Molecular mechanisms of proteasome assembly. *Nature reviews molecular cell biology* 2009;10:104–15.

40. Tanaka, K, Mizushima, T, Saeki, Y The proteasome: molecular machinery and pathophysiological roles. *Biological chemistry* 2012;393:217–34.

41. Mannhaupt, G, Schnall, R, Karpov, V, Vetter, I, Feldmann, H Rpn4p acts as a transcription factor by binding to PACE, a nonamer box found upstream of 26S proteasomal and other genes in yeast. *FEBS letters* 1999;450:27–34.

42. Chapple, SJ, Siowand, RCM, Mann, GE Crosstalk between Nrf2 and the proteasome: therapeutic potential of Nrf2 inducers in vascular disease and aging. *The international journal of biochemistry & cell biology* 2012;44:1315–20.

43. Steffen, J, Seeger, M, Koch, A, Krüger, E Proteasomal degradation is transcriptionally controlled by TCF11 via an ERAD-dependent feedback loop. *Molecular cell* 2010;40:147–58.

44. Bedford, L, Lowe, J, Dick, LR, Mayer, RJ, Brownell, JE Ubiquitin-like protein conjugation and the ubiquitin-proteasome system as drug targets. *Nature reviews drug discovery* 2011;10:29–46.

45. Schubert, U, Antón, LC, Gibbs, J, Norbury, CC, Yewdell, JW, Bennink, JR Rapid degradation of a large fraction of newly synthesized proteins by proteasomes. *Nature* 2000;404:770–4.

46. Wickner, S Posttranslational quality control: folding, refolding, and degrading proteins. *Science* 1999;286:1888–93.

47. Rosser, MFN, Washburn, E, Muchowski, PJ, Patterson, C, Cyr, DM Chaperone functions of the E3 ubiquitin ligase CHIP. *The journal of biological chemistry* 2007;282:22267–77.

48. Min, J-N, Whaley, RA, Sharpless, NE, Lockyer, P, Portbury, AL, Patterson, C CHIP deficiency decreases longevity, with accelerated aging phenotypes accompanied by altered protein quality control. *Molecular and cellular biology* 2008;28:4018–25.

49. Dickey, CA, Kamal, A, Lundgren, K, et al. The high-affinity HSP90-CHIP complex recognizes and selectively degrades phosphorylated tau client proteins. *Journal of clinical investigations* 2007;117:648–58.

50. Chhangani, D, Joshi, AP, Mishra, A E3 ubiquitin ligases in protein quality control mechanism. *Molecular neurobiology* 2012;45:571–85.

51. Farout, L, Friguet, B Proteasome function in aging and oxidative stress: implications in protein maintenance failure. *Antioxidants & redox signaling* 2006;8:205–16.

52. Jung, T, Grune, T The proteasome and its role in the degradation of oxidized proteins. *IUBMB life* 2008;60:743–52.

53. Dasuri, K, Zhang, L, Ebenezer, P, Liu, Y, Fernandez-Kim, SO, Keller, JN Aging and dietary restriction alter proteasome biogenesis and composition in the brain and liver. *Mechanisms of ageing and development* 2009;130:777–83.

54. Keller, JN, Gee, J, Ding, Q The proteasome in brain aging. *Ageing research reviews* 2002;1:279–93.

55. Chondrogianni, N, Gonos, ES Proteasome dysfunction in mammalian aging: steps and factors involved. *Experimental gerontology* 2005;40:931–8.

56. Chondrogianni, N, Petropoulos, I, Franceschi, C, Friguet, B, Gonos, ES Fibroblast cultures from healthy centenarians have an active proteasome. *Experimental gerontology* 2000;35:721–8.

57. Yun, C, Stanhill, A, Yang, Y, et al. Proteasomal adaptation to environmental stress links resistance to proteotoxicity with longevity in *Caenorhabditis elegans*. *Proceedings of the national academy of sciences of the united states of America* 2008;105:7094–9.

58. Tonoki, A, Kuranaga, E, Tomioka, T, et al. Genetic evidence linking age-dependent attenuation of the 26S proteasome with the aging process. *Molecular and cellular biology* 2009;29:1095–106.

59. Vernace, VA, Arnaud, L, Schmidt-Glenewinkel, T, Figueiredo-Pereira, ME Aging perturbs 26S proteasome assembly in *Drosophila melanogaster*. *FASEB journal* 2012;21:2672–82.

60. Vernace, VA, Schmidt-Glenewinkel, T, Figueiredo-Pereira, ME Aging and regulated protein degradation: who has the UPPer hand? *Aging cell* 2007;6:599–606.

61. Carrard, G, Dieu, M, Raes, M, Toussaint, O, Friguet, B Impact of ageing on proteasome structure and function in human lymphocytes. *The international journal of biochemistry & cell biology* 2003;35:728–39.

62. Bence, NF, Sampat, RM, Kopito, RR Impairment of the ubiquitin-proteasome system by protein aggregation. *Science* 2001;292:1552–5.

63. Chondrogianni, N, Tzavelas, C, Pemberton, AJ, Nezis, IP, Rivett, AJ, Gonos, ES Overexpression of proteasome beta5 assembled subunit increases the amount of proteasome and confers ameliorated response to oxidative stress and higher survival rates. *The journal of biological chemistry* 2005;280:11840–50.

64. Chondrogianni, N, Stratford, FLL, Trougakos, IP, Friguet, B, Rivett, AJ, Gonos, ES Central role of the proteasome in senescence and survival of human fibroblasts: induction of a senescence-like phenotype upon its inhibition and resistance to stress upon its activation. *The journal of biological chemistry* 2003;278:28026–37.

65. Vilchez, D, Morantte, I, Liu, Z, et al. RPN-6 determines *C. elegans* longevity under proteotoxic stress conditions. *Nature* 2012;489:263–8.

66. Carrano, AC, Liu, Z, Dillin, A, Hunter, T A conserved ubiquitination pathway determines longevity in response to diet restriction. *Nature* 2010;460:396–9.

67. Zhang, H, Cohen, SN Smurf2 up-regulation activates telomere-dependent senescence. *Genes & development* 2004;18:3028–40.

68. Ramkumar, C, Kong, Y, Cui, H, et al. Smurf2 regulates the senescence response and suppresses tumorigenesis in mice. *Cancer research* 2012;72:2714–9.

69. Byrne, A, McLaren, RP, Mason, P, et al. Knockdown of human deubiquitinase PSMD14 induces cell cycle arrest and senescence. *Experimental cell research* 2010;316:258–71.

70. Mijaljica, D, Prescott, M, Devenish, RJ The intricacy of nuclear membrane dynamics during nucleophagy. *Nucleus* 2010;1:213–23.

71. Rambold, AS, Lippincott-Schwartz, J Mechanisms of mitochondria and autophagy crosstalk. *Cell cycle* 2011;10:4032–8.

72. Claessen, JHL, Kundrat, L, Ploegh, HL Protein quality control in the ER: balancing the ubiquitin checkbook. *Trends in cell biology* 2012;22:22–32.

73. Perlmutter, DH Alpha-1-antitrypsin deficiency: importance of proteasomal and autophagic degradative pathways in disposal of liver disease-associated protein aggregates. *Annual review of medicine* 2011;62:333–45.

74. Hirsch, C, Gauss, R, Horn, SC, Neuber, O, Sommer, T The ubiquitylation machinery of the endoplasmic reticulum. *Nature* 2009;458:453–60.

75. Ron, D, Walter, P Signal integration in the endoplasmic reticulum unfolded protein response. *Nature reviews molecular cell biology* 2007;8:519–29.

76. Wang, S, Kaufman, RJ The impact of the unfolded protein response on human disease. *The journal of cell biology* 2012;197:857–67.

77. Shore, GC, Papa, FR, Oakes, SA Signaling cell death from the endoplasmic reticulum stress response. *Current opinion in cell biology* 2011;23:143–9.

78. McElvaney, NG, Greene, CM Mechanisms of protein misfolding in conformational lung diseases. *Current molecular medicine* 2012;12:850–9.

79. Brown, MK, Naidoo, N The endoplasmic reticulum stress response in aging and age-related diseases. *Frontiers in physiology* 2012;3:263.

80. Torres-González, E, Bueno, M, Tanaka, A, et al. Role of endoplasmic reticulum stress in age-related susceptibility to lung fibrosis. *American journal of respiratory cell and molecular biology* 2012;46:748–756.

81. Sadighi Akha, AA, Harper, JM, Salmon, AB, et al. Heightened induction of proapoptotic signals in response to endoplasmic reticulum stress in primary fibroblasts from a mouse model of longevity. *The journal of biological chemistry* 2011;286:30344–51.

82. Henis-korenblit, S, Zhang, P, Hansen, M, et al. Insulin/IGF-1 signaling mutants reprogram ER stress response regulators to promote longevity. *Proceedings of the national academy of sciences of the united states of America* 2010;107:9730–5.

83. Appenzeller-Herzog, C, Hall, MN Bidirectional crosstalk between endoplasmic reticulum stress and mTOR signaling. *Trends in cell biology* 2012;22:274–82.

84. Wong, E, Cuervo, AM Integration of clearance mechanisms: the proteasome and autophagy. *Cold spring harbor perspectives in biology* 2010;2:a006734.

85. Dahlmann, B Role of proteasomes in disease. *BMC biochemistry* 2007;8 Suppl 1:S3.

86. Jones, DL, Rando, TA Emerging models and paradigms for stem cell ageing. *Nature cell biology* 2011;13:506–12.

87. Ponnappan, S, Ovaa, H, Ponnappan, U Lower expression of catalytic and structural subunits of the proteasome contributes to decreased proteolysis in peripheral blood T lymphocytes during aging. *The international journal of biochemistry & cell biology* 2007;39:799–809.

88. Kirkwood, TBL, Austad, SN Why do we age? 2000;408:233–8.

89. Crotzer, VL, Blum, JS Autophagy and adaptive immunity. *Immunology* 2010;131:9–17.

90. Calamini, B, Silva, MC, Madoux, F, et al. Small-molecule proteostasis regulators for protein conformational diseases. *Nature chemical biology* 2012;8:185–96.

4 Telomerase Function in Aging

Rodrigo T. Calado

University of São Paulo at Ribeirão, Preto Medical School, São Paulo, Brazil

4.1 Telomeres

The evolution from circular chromosomes in bacteria to linear chromosomes in higher species posed a problem to the genome structure by creating **natural** ends at the DNA molecules that needed to be distinguished from broken DNA. Telomeres are the structures formed at the extremities (**natural ends**) of linear chromosomes that serve to protect them from damage and from the inevitable shortening of DNA due to asymmetrical DNA replication (Figure 4.1) [1]. For DNA synthesis, the DNA polymerase requires a 3′ hydroxyl donor group from a nucleotide to initiate DNA replication, usually provided by an RNA primer; after the DNA polymerase moves along the DNA single strand, the RNA primer detaches, but the empty space left by the primer is filled by work of another DNA polymerase in the adjacent Okazaki fragment, and the entire DNA strand is duplicated. However, at the very end of the linear DNA, there is no adjacent Okazaki fragment or DNA polymerase, and the 3′ end originally filled with the RNA primer remains unduplicated; this phenomenon was acknowledged early in molecular biology and named the **end-replication problem** or **marginotomy** [2]. As a consequence, the newly synthesized DNA strand is shorter than the original template and chromosomes become shorter after every mitotic division. This imposes a problem to the cell as it divides: if the cell experiences too many mitotic divisions, it may start losing genomic material and genetic information. In all vertebrate cells, telomeres are composed of repetitive hexameric DNA sequences (TTAGGG in the leading strand and CCCTAA in the lagging strand) coated by specific DNA-binding proteins, collectively called shelterin [3]. Some of these proteins recognize and directly bind to double-stranded telomeric DNA (TRF1 and TRF2), while POT1 binds to single-stranded telomeric DNA, and TPP1 and RAP1 indirectly binds to DNA through other proteins (Figure 4.1b). The telomere DNA sequence ends as a single-strand 3′ overhang that folds back and binds to upstream telomeric DNA, forming a lariat structure at the very ends of chromosomes like a knot: the T-loop (Figure 4.1). This complex structure of the DNA strands combined with telomere-binding proteins protects and hides the natural ends of chromosomes.

Molecular Aspects of Aging: Understanding Lung Aging, First Edition. Edited by Mauricio Rojas, Silke Meiners and Claude Jourdan Le Saux.

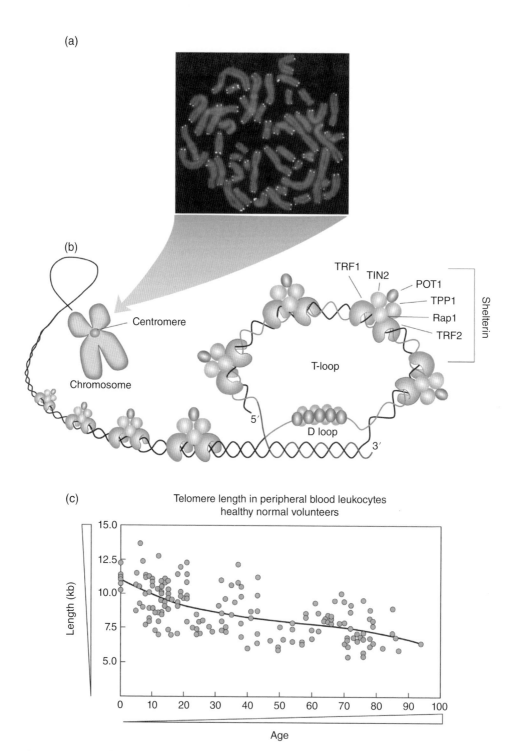

(a)

(b)

TRF1
TIN2
POT1
TPP1
Rap1
TRF2
Shelterin

Centromere

Chromosome

T-loop

5′

D loop

3′

(c)

Telomere length in peripheral blood leukocytes
healthy normal volunteers

Length (kb)

15.0

12.5

10.0

7.5

5.0

0   10   20   30   40   50   60   70   80   90   100

Age

Figure 4.1   (a) Telomeres are ribonucleoprotein structures at the ends of linear chromosomes and can be detected by fluorescence *in situ* hybridization (FISH, yellow bright signals) and the extremities of each chromatide of the chromosomes (blue structures) during metaphase. (b) Schematically, telomeres are composed of hundreds to thousands of TTAGGG hexameric DNA repeats coated by specialized proteins collectively termed shelterin, forming a lariat at the very end of the molecule (T-loop). (c) Telomeres shorten with human aging. The graphic represents the lengths of telomeres of peripheral blood leukocytes from birth to 100 years in healthy volunteers. Telomeres are measured in kilobases (kb) and each circle represents one individual. To see a color version of this figure, see Plate 4.1.

Telomeres function as a **buffer** at the chromosome ends and mitosis causes the loss of telomeric repeats, avoiding the deletion of genetically encoded information and alleviating the deleterious effects of the **end-replication problem** to the genome. Telomere shortening is inevitable as cells divide and the organism ages. When telomeres reach critically short lengths, they activate intracellular signaling pathways, especially the p53 tumor suppressor protein, thus inhibiting cell proliferation and inducing senescence and/or apoptosis [4]. In human peripheral blood white cells, for instance, telomeres consist of hundreds to thousands of TTAGGG repeats, varying from 10–12 kilobases in the newborn to 5–6 kilobases in the elderly (Figure 4.1c) [5].

Telomeres provide the molecular mechanism for the limited proliferation capacity of cells either *in vivo* or *in vitro*, the latter termed the **Hayflick limit** [6]. Leonard Hayflick observed in the 1960s that human adult cells do not proliferate indefinitely, that is, are not immortal, but cease dividing after a certain number of mitotic divisions. In a Petri dish, human fetal cells divide up to 60 times and then engage in proliferative senescence. Hayflick also determined that this proliferation arrest is due to mechanisms intrinsic to the cell rather than inhibitors or **toxins** secreted into the culture medium. In fact, telomeres serve as a **mitotic clock** and telomere shortening is considered to be responsible for the Hayflick limit in adult mature cells. Replicative senescent cells have a flattened cytoplasm and increased volume, altered gene expression profile, and abnormal nuclear structure and protein processing but remain metabolically viable [7].

4.2 Telomerase

In contrast to mature differentiated cells, some cells require high proliferative capacity, such as embryonic and tissue-specific adult stem cells and T- and B-lymphocytes. To circumvent the telomere erosion resultant from high proliferation, these cells maintain telomere length using an unusual mechanism for DNA synthesis in vertebrate cells: reverse transcription. These cells express telomerase (telomerase reverse transcriptase (TERT)), a reverse transcriptase (RT) enzyme that uses a single-stranded RNA molecule (TERC) as a template to add TTAGGG hexameric repeats to the 3' end of the DNA leading strand (Figure 4.2). Telomerase is a long protein (1132 amino acids) with at least three domains: N-terminal, RT motifs, and C-terminal. Other proteins associate with telomerase and are part of the telomerase holoenzyme: dyskerin (encoded by the X-linked *DKC1* gene), NOP10 (H/ACA small nucleolar protein family A), NHP2 (nonhistone protein 2, also an H/ACA small nucleolar protein), and GAR1 (glycine/arginine-rich, H/ACA small nucleolar protein homolog). In addition, TCAB1 (encoded by *WRAP53*) is an essential component of the telomerase complex and serves to localize telomerase to the Cajal bodies for telomere elongation [8, 9].

That telomere erosion is the molecular event responsible for the Hayflick limit is supported by telomerase expression. When telomerase is ectopically expressed in mature cells, such as fibroblasts, by, for instance, gene transfection, the transfected cells overcome the Hayflick limit and reach expanded proliferation capabilities [10]. It is important to note, however, that telomerase expression alone does not lead to a **transformed phenotype**, nor is sufficient to induce malignant transformation [11].

Telomerase expression is tightly regulated and most mature human cells do not express telomerase [12]. The expression of the components of telomerase holoenzyme is controlled at transcription, alternative splicing, assembly, subcellular localization, and posttranslational levels. At the transcription level, the telomerase gene may be highly methylated. The retinoblastoma

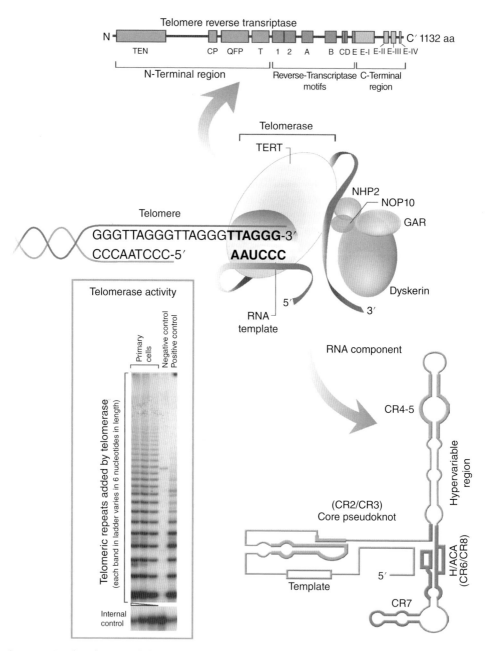

Figure 4.2 The telomerase holoenzyme is composed of the catalytic unit, TERT, its RNA component (TERC) that serves as a template for telomere elongation, and associated proteins (dyskerin, GAR, NOP10, and NHP2). The complex attaches to the 3′ end of the telomeric DNA and adds hexameric repeats. The inset shows a method to evaluate telomerase activity. Telomerase enzymatic activity may be measured *in vitro* by PCR amplification of telomerase products that are run in a gel. Each band represents the telomerase product and is separated by six nucleotides. To see a color version of this figure, see Plate 4.2.

protein (Rb) and p21 suppress its expression, whereas MYC, the Wnt/β-catenin pathway, KLF4, and estrogens stimulate its transcription by acting on the promoter region [5, 13]. *TERT* has several splicing variants, some of which have a dominant-negative effect over telomerase enzymatic activity [14]. Dyskerin, NOP10, NHP2, and GAR1 are responsible for the holoenzyme assembly and complex stability, and TCAB1 (telomerase Cajal body protein 1) has pivotal role in the complex trafficking and drives it close to telomeres to exert its function. On the other hand, tripeptidyl peptidase I (TPP1), a sheltering component, inhibits telomerase access to telomeres, thus preventing telomere elongation [15].

As mentioned above, telomerase is mainly expressed in embryonic and adult stem cells, which demand high proliferative function. The highest levels of telomerase expression are observed in embryonic stem cells, which translates into maintenance of telomere length in spite of consecutive cell division either *in vivo* or *in vitro* [16, 17, 18]. As these cells differentiate during development, telomerase expression is downregulated, but adult tissue-specific stem cells retain some telomerase expression [12]. However, telomerase expression levels in adult stem cells are not sufficient to maintain telomere lengths, and the net balance between elongation and erosion results in continuous attrition of telomeres with organism aging, as observed in the hematopoietic tissue (Figure 4.1c) [12]. Although hematopoietic stem cells express telomerase, its levels are not satisfactory for maintaining the same length of telomeres over time [19].

4.3 Telomeres and human disease

Telomerase-deficient murine models show progressive telomere shortening over generations and eventually become infertile after five to six generations [20, 21]. Telomeres eventually become dysfunctional, and chromosome instability and aneuploidy occur. Different tissues display features of proliferative senescence; the bone marrow has reduced numbers of hematopoietic stem and progenitor cells, recovery after chemotherapy is abrogated, and the gut shows villus atrophy. When combined with a p53 defect, telomerase-deficient mice also display increased epithelial cancer susceptibility [22].

These features somehow resemble the characteristics of patients in which telomerase is inherently deficient [23]. The prototypical disease of telomere shortening is dyskeratosis congenita, an inherited disorder characterized by mucocutaneous dystrophy and bone marrow failure (aplastic anemia). Typical cases of dyskeratosis congenita are X-linked, and mutations in the *DKC1* gene, which encodes dyskerin, an essential component of the telomerase holoenzyme, are etiologic [24]. In these patients, as mutations are hemizygous, telomerase activity is severely reduced or virtually absent [25]. Dyskeratosis congenita also may occur when both alleles for a telomerase component are mutated: *TERT*, *TERC*, *NOP10*, or *TCAB1*. In addition, mutations in the shelterin component *TINF2* lead to severe telomere dysfunction and shortening due to abnormal DNA capping and clinically manifest as dyskeratosis congenita. Most patients with dyskeratosis congenita die from bone marrow failure, and when aplastic anemia is rescued by hematopoietic stem cell transplant, patients may succumb from cancer development. In addition, some patients with dyskeratosis congenita also develop liver and lung disorders.

Less pleiotropic manifestations of telomerase deficiency also may occur. When monoallelic mutations are found in *TERT* or *TERC* (with a remaining wild-type allele) and telomerase function is reduced to about 50% of normal, the clinical phenotype usually affects a single organ. Some patients with apparently acquired aplastic anemia, a commonly immune-mediated disorder, may be heterozygous for *TERT* or *TERC* loss-of-function mutations and display very short telomeres in hematopoietic cells, low telomerase activity, reduced hematopoietic stem cell pool, and an empty bone marrow [26, 27, 28, 29]. However, these patients do not present the classical features of dyskeratosis congenita, namely, the skin abnormalities or lung or liver dysfunction.

Other patients heterozygous for mutations in the same genes, however, may have different clinical presentation and other organs affected. *TERT* or *TERC* mutations are the most common genetic risk factor for idiopathic pulmonary fibrosis [30, 31] (discussed below), and telomerase mutations causing telomere erosion also are seen in patients with hepatic cirrhosis, either cryptic [32] or associated with alcohol consumption or hepatitis viruses [33, 34].

The current understanding is that in these patients, there is a strong association between the genetic events (telomerase deficiency, telomere shortening) and environmental factors [5]. Telomere attrition may induce deficient organ regeneration, and in the event of an environmental challenge, such as smoking for the lungs or alcohol and viruses for the liver, tissue repair is abrogated by deficient telomerase function necessary for cell proliferation and tissue regeneration, thus evolving to organ damage.

Patients with telomerase mutations and telomere shortening also have a proclivity for cancer development. Dyskeratosis congenita patients have a 1000-fold chance of developing tongue squamous cell carcinoma and a 200-fold chance of developing leukemia in comparison to age-matched healthy subjects [35]. Constitutional telomerase mutations also are risk factors for developing acute myeloid leukemia [36], and the *TERT* gene is a susceptibility locus for a variety of cancers [37]. Additionally, in population studies, individuals with shorter telomeres have a much higher chance of developing cancer [38]. Telomere erosion reduces chromosome stability and precipitates the appearance of aneuploidy, translocations, and point mutations both in mice and humans [22, 39]. This appears to be an important early step in the progression of a normal cell to a malignant phenotype.

4.3.1 Telomere dysfunction in the lungs

Approximately 20% of patients with dyskeratosis congenita eventually develop pulmonary fibrosis [40]. This complication is more common after bone marrow transplant, when patients undergo intensive chemotherapy to ablate the marrow, but some degree of lung toxicity is expected. However, dyskeratosis congenita patients are unusually more susceptible to these complications and often die from respiratory failure early after transplant [41].

Three separate studies using different approaches identified heterozygous mutations in telomerase genes in both familial and sporadic idiopathic pulmonary fibrosis patients [30, 31, 42]. Most patients with telomerase mutations and pulmonary involvement have usual interstitial pneumonia histologic pattern [31]. In parallel to what happens in the liver of patients with short telomeres, pulmonary fibrosis also was described as cirrhosis of the lungs almost 200 years ago, with "contracting fibres...obliterating its small air tubes and its blood vessels" [43, 44]. However, why telomere dysfunction eventually causes pulmonary fibrosis is not well understood. Almost all patients with idiopathic pulmonary fibrosis have short telomeres, regardless of their mutations status [45]. In addition, pulmonary fibrosis is most prevalent at older age, and telomeres shorten with aging (Figure 4.1). Smoking, an important fibrogenic risk factor, also appears to induce telomere shortening [46]. Taken together, these observations suggest that telomere erosion, whether determined by genetics, environment, or aging, contributes to lung fibrosis. Available laboratory evidence suggests that idiopathic pulmonary fibrosis may be a disease of defective regeneration of the lungs associated with local inflammation [30]. Telomere shortening may inhibit proliferation of pneumocytes type II in the lungs and halt adequate regeneration during or after exposure to toxins [30]. However, it is not clear how telomere attrition may act on lung-resident mesenchymal cells involved in fibrosis formation. In a family with idiopathic pulmonary fibrosis, family members were studied 20 years before developing pulmonary fibrosis, and in their early bronchoalveolar lavages, inflammatory infiltrate was mainly composed of macrophages, in spite of normal CT scans, suggesting that telomerase deficiency also engages an early abnormal inflammatory response in the lungs [47].

4.4 Telomeres biology, aging, and longevity

Telomeres inevitably shorten with aging, since telomere erosion is faster than telomere elongation in adult stem cells (Figure 4.2). Consequently, tissues with higher proliferative demands, such as the bone marrow and the skin, may exhibit more obvious telomere attrition in their mature cells with aging. In addition, there is a wide variability in telomere length for individuals with the same age (Figure 4.1), reflecting the effects of additional genetic determinants and environmental agents on telomere dynamics. There are genetic factors that contribute to telomere length [48] with stronger maternal inheritance, but also paternal age at conception appears to modulate the length of telomeres in the offspring [48, 49]. Telomere lengths are different in different cell types from the same individual [50]. For instance, telomeres are shorter in peripheral blood leukocytes than in skin fibroblasts from the same person, which may be anticipated by the more active mitotic history of blood cells relative to skin fibroblasts. Even within blood white cells, telomeres are heterogeneous among subtypes; telomeres are shorter in lymphocytes as compared to neutrophils, and T regulatory cells tend to have the shortest telomeres among lymphocyte subsets [12]. Telomeres also shorten at different rates with aging in different cells. Telomere erosion with age is faster in lymphocytes in comparison to granulocytes [12]. And even within the same cell, telomeres are heterogeneous in length when chromosomes are compared; chromosome 17p consistently has the shortest telomeres in humans [51]. Finally, external factors also may modulate the rate of telomere attrition with aging. Smoking, diet, and social economic status have been blamed for faster telomere shortening [52]. At the molecular level, reactive oxygen species damage DNA and exposure to these radicals may accelerate telomere shortening *in vitro* [53]. When cell lines are cultured at low oxygen concentration (2–5%) conditions, telomere erosion is alleviated as compared to cells cultured at room air (21% oxygen) conditions [54]. In summary, telomere length is genetically determined at birth with wide variability, but exposure to environmental agents during life modulates the rate of telomere attrition as we age.

However, it is still unclear whether telomere shortening is a cause or a consequence of aging. Telomere length is the biomarker that best correlates with aging in a given subject. As mentioned earlier, telomere attrition is a key molecular element responsible for cellular replicative senescence and produces metabolic and morphologic changes in the cell, although they may remain viable. This does not necessarily translate, however, into organism aging. That telomere shortening may be an active player in aging was raised by the observation of short telomeres in **Dolly** the sheep, suggesting that nuclear transferring and reprograming are not able to restore telomeres [55]. However, others have shown that nuclear transfer produces animals with longer telomeres and increased lifespan [56]. In fact, embryonic stem cells express high levels of telomerase, and reprograming mature cells to pluripotency state by ectopic expression of transcription factors (iPS cells) also leads to telomere elongation [16, 17, 18].

Dyskeratosis congenita is not a progeroid syndrome; patients do not look older than their chronological age and do not develop diseases commonly associated with age – cardiovascular disease or neurodegenerative disorders [5]. Aplastic anemia is not a disorder of the elderly, but is mainly seen in the first decades of life [57]. Interestingly, laboratory mouse telomeres are 5 to 10 times longer than in humans, but they age much faster than humans do and their lifespan is 30 times shorter [58].

However, patients with telomerase deficiency have a higher susceptibility for cancer development, and pulmonary fibrosis and cirrhosis are more common at older ages. More importantly, their adult stem cells are **aged** and depleted and they show abnormal regeneration properties, similar to what is observed in the elderly.

Although the cause–effect relationship between telomere erosion and aging is still to be better understood, there is a clearer connection between telomeres and longevity, at least within humans. While aging is a process, longevity is a measure and a longer lifespan does not necessarily implicate a slower aging process. On one end of the spectrum, patients with telomerase defects have severe diseases early in life and eventually succumb from marrow failure, lung fibrosis, cirrhosis, or cancer. On the other end, in the general population, telomere length appears to predict survival. In one study of individuals older than 60 years, telomere length inversely correlated with survival, and the increased causes of deaths were infection and cardiovascular diseases [59]. However, these findings were not replicated by some other studies [60, 61]. Finally, in an Italian cohort, patients with shorter telomeres had a higher probability of developing cancer, and cancer mortality also inversely correlated with telomere length [38].

4.5 Conclusion

Telomeres are special structures at the ends of chromosomes responsible for genome stability, and telomere erosion is a fundamental biologic process responsible for cell replicative senescence. Telomerase alleviates telomere erosion due to mitotic cell division, and telomerase defects are associated with human diseases – bone marrow failure, pulmonary fibrosis, cirrhosis, and cancer. Although telomeres shorten with aging, it does not appear to directly drive aging itself, but telomere length may be a biomarker for longevity in humans.

References

1. Blackburn, E.H., *Switching and signaling at the telomere. Cell*, 2001. **106**(6): p. 661–73.
2. Olovnikov, A.M., *Principle of marginotomy in template synthesis of polynucleotides. Dokl Akad Nauk SSSR*, 1971. **201**(6): p. 1496–9.

3. de Lange, T., *How shelterin solves the telomere end-protection problem. Cold Spring Harb Symp Quant Biol*, 2010. **75**: p. 167–77.

4. Collado, M., M.A. Blasco, and M. Serrano, *Cellular senescence in cancer and aging. Cell*, 2007. **130**(2): p. 223–33.

5. Calado, R.T. and N.S. Young, *Telomere diseases. N Engl J Med*, 2009. **361**(24): p. 2353–65.

6. Hayflick, L. and P.S. Moorhead, *The serial cultivation of human diploid cell strains. Exp Cell Res*, 1961. **25**: p. 585–621.

7. Ben-Porath, I. and R.A. Weinberg, *When cells get stressed: an integrative view of cellular senescence. J Clin Invest*, 2004. **113**(1): p. 8–13.

8. Venteicher, A.S., E.B. Abreu, Z. Meng, et al., *A human telomerase holoenzyme protein required for Cajal body localization and telomere synthesis. Science*, 2009. **323**(5914): p. 644–8.

9. Venteicher, A.S. and S.E. Artandi, *TCAB1: driving telomerase to Cajal bodies. Cell Cycle*, 2009. **8**(9): p. 1329–31.

10. Bodnar, A.G., M. Ouellette, M. Frolkis, et al., *Extension of life-span by introduction of telomerase into normal human cells. Science*, 1998. **279**(5349): p. 349–52.

11. Hahn, W.C., C.M. Counter, A.S. Lundberg, R.L. Beijersbergen, M.W. Brooks, and R.A. Weinberg, *Creation of human tumour cells with defined genetic elements. Nature*, 1999. **400**(6743): p. 464–8.

12. Aubert, G. and P.M. Lansdorp, *Telomeres and aging. Physiol Rev*, 2008. **88**(2): p. 557–79.

13. Hoffmeyer, K., A. Raggioli, S. Rudloff, et al., *Wnt/beta-catenin signaling regulates telomerase in stem cells and cancer cells. Science*, 2012. **336**(6088): p. 1549–54.

14. Colgin, L.M., C. Wilkinson, A. Englezou, A. Kilian, M.O. Robinson, and R.R. Reddel, *The hTERTalpha splice variant is a dominant negative inhibitor of telomerase activity. Neoplasia*, 2000. **2**(5): p. 426–32.

15. Zhong, F.L., L.F. Batista, A. Freund, M.F. Pech, A.S. Venteicher, and S.E. Artandi, *TPP1 OB-fold domain controls telomere maintenance by recruiting telomerase to chromosome ends. Cell*, 2012. **150**(3): p. 481–94.

16. Thomson, J.A., J. Itskovitz-Eldor, S.S. Shapiro, et al., *Embryonic stem cell lines derived from human blastocysts. Science*, 1998. **282**(5391): p. 1145–7.

17. Marion, R.M., K. Strati, H. Li, et al., *Telomeres acquire embryonic stem cell characteristics in induced pluripotent stem cells. Cell Stem Cell*, 2009. **4**(2): p. 141–54.

18. Winkler, T., S.G. Hong, J.E. Decker, et al., *Defective telomere elongation and hematopoiesis from telomerase-mutant aplastic anemia iPSCs. J Clin Invest*. 2013. **123**(5): p. 1952–63.

19. Wang, J.C., J.K. Warner, N. Erdmann, P.M. Lansdorp, L. Harrington, and J.E. Dick, *Dissociation of telomerase activity and telomere length maintenance in primitive human hematopoietic cells. Proc Natl Acad Sci USA*, 2005. **102**(40): p. 14398–403.

20. Blasco, M.A., H.W. Lee, M.P. Hande, et al., *Telomere shortening and tumor formation by mouse cells lacking telomerase RNA. Cell*, 1997. **91**(1): p. 25–34.

21. Chiang, Y.J., R.T. Calado, K.S. Hathcock, P.M. Lansdorp, N.S. Young, and R.J. Hodes, *Telomere length is inherited with resetting of the telomere set-point. Proc Natl Acad Sci USA*, 2010. **107**(22): p. 10148–53.

22. Artandi, S.E., S. Chang, S.L. Lee, et al., *Telomere dysfunction promotes non-reciprocal translocations and epithelial cancers in mice. Nature*, 2000. **406**(6796): p. 641–5.

23. Calado, R.T., *Telomeres and marrow failure. Hematology Am Soc Hematol Educ Program*, 2009: p. 338–43.

24. Heiss, N.S., S.W. Knight, T.J. Vulliamy, et al., *X-linked dyskeratosis congenita is caused by mutations in a highly conserved gene with putative nucleolar functions. Nat Genet*, 1998. **19**(1): p. 32–8.

25. Knight, S.W., N.S. Heiss, T.J. Vulliamy, et al., *X-linked dyskeratosis congenita is predominantly caused by missense mutations in the DKC1 gene. Am J Hum Genet*, 1999. **65**(1): p. 50–8.

26. Yamaguchi, H., R.T. Calado, H. Ly, et al., *Mutations in TERT, the gene for telomerase reverse transcriptase, in aplastic anemia. N Engl J Med*, 2005. **352**(14): p. 1413–24.

27. Fogarty, P.F., H. Yamaguchi, A. Wiestner, et al., *Late presentation of dyskeratosis congenita as apparently acquired aplastic anaemia due to mutations in telomerase RNA. Lancet*, 2003. **362**(9396): p. 1628–30.

28. Ly, H., R.T. Calado, P. Allard, et al., *Functional characterization of telomerase RNA variants found in patients with hematologic disorders. Blood*, 2005. **105**(6): p. 2332–9.

29. Xin, Z.T., A.D. Beauchamp, R.T. Calado, J.W. et al., *Functional characterization of natural telomerase mutations found in patients with hematologic disorders. Blood*, 2007. **109**(2): p. 524–32.

30. Armanios, M.Y., J.J. Chen, J.D. Cogan, et al., *Telomerase mutations in families with idiopathic pulmonary fibrosis. N Engl J Med*, 2007. **356**(13): p. 1317–26.

31. Tsakiri, K.D., J.T. Cronkhite, P.J. Kuan, et al., *Adult-onset pulmonary fibrosis caused by mutations in telomerase. Proc Natl Acad Sci USA*, 2007. **104**(18): p. 7552–7.

32. Calado, R.T., J.A. Regal, D.E. Kleiner, et al., *A spectrum of severe familial liver disorders associate with telomerase mutations. PLoS One*, 2009. **4**(11): p. e7926.

33. Calado, R.T., J. Brudno, P. Mehta, et al., *Constitutional telomerase mutations are genetic risk factors for cirrhosis. Hepatology*, 2011. **53**(5): p. 1600–7.

34. Hartmann, D., U. Srivastava, M. Thaler, et al., *Telomerase gene mutations are associated with cirrhosis formation. Hepatology*, 2011. **53**(5): p. 1608–17.

35. Alter, B.P., N. Giri, S.A. Savage, and P.S. Rosenberg, *Cancer in dyskeratosis congenita. Blood*, 2009. **113**(26): p. 6549–57.

36. Calado, R.T., J.A. Regal, M. Hills, et al., *Constitutional hypomorphic telomerase mutations in patients with acute myeloid leukemia. Proc Natl Acad Sci USA*, 2009. **106**(4): p. 1187–92.

37. Rafnar, T., P. Sulem, S.N. Stacey, et al., *Sequence variants at the TERT-CLPTM1L locus associate with many cancer types. Nat Genet*, 2009. **41**(2): p. 221–7.

38. Willeit, P., J. Willeit, A. Mayr, et al., *Telomere length and risk of incident cancer and cancer mortality. JAMA*, 2010. **304**(1): p. 69–75.

39. Calado, R.T., J.N. Cooper, H.M. Padilla-Nash, et al., *Short telomeres result in chromosomal instability in hematopoietic cells and precede malignant evolution in human aplastic anemia. Leukemia*, 2012. **26**(4): p. 700–7.

40. Dokal, I., *Dyskeratosis congenita in all its forms. Br J Haematol*, 2000. **110**(4): p. 768–79.

41. Rocha, V., A. Devergie, G. Socie, et al., *Unusual complications after bone marrow transplantation for dyskeratosis congenita. Br J Haematol*, 1998. **103**(1): p. 243–8.

42. Mushiroda, T., S. Wattanapokayakit, A. Takahashi, et al., *A genome-wide association study identifies an association of a common variant in TERT with susceptibility to idiopathic pulmonary fibrosis. J Med Genet*, 2008. **45**(10): p. 654–6.

43. Corrigan, D.J., *On cirrhosis of the lung. Dublin J Med Sci*, 1838. **13**: p. 20.

44. Thannickal, V.J. and J.E. Loyd, *Idiopathic pulmonary fibrosis: a disorder of lung regeneration? Am J Respir Crit Care Med*, 2008. **178**(7): p. 663–5.

45. Alder, J.K., J.J. Chen, L. Lancaster, et al., *Short telomeres are a risk factor for idiopathic pulmonary fibrosis. Proc Natl Acad Sci USA*, 2008. **105**(35): p. 13051–6.

46. Morla, M., X. Busquets, J. Pons, J. Sauleda, W. MacNee, and A.G. Agusti, *Telomere shortening in smokers with and without COPD. Eur Respir J*, 2006. **27**(3): p. 525–8.

47. El-Chemaly, S., S.G. Ziegler, R.T. Calado, et al., *Natural history of pulmonary fibrosis in two subjects with the same telomerase mutation. Chest*, 2011. **139**(5): p. 1203–9.

48. Graakjaer, J., H. Der-Sarkissian, A. Schmitz, et al., *Allele-specific relative telomere lengths are inherited. Hum Genet*, 2006. **119**(3): p. 344–50.

49. Broer, L., V. Codd, D.R. Nyholt, et al., *Meta-analysis of telomere length in 19 713 subjects reveals high heritability, stronger maternal inheritance and a paternal age effect. Eur J Hum Genet*, 2013.

50. Gadalla, S.M., R. Cawthon, N. Giri, B.P. Alter, and S.A. Savage, *Telomere length in blood, buccal cells, and fibroblasts from patients with inherited bone marrow failure syndromes. Aging*, 2010. **2**(11): p. 867–74.

51. Martens, U.M., J.M. Zijlmans, S.S. Poon, et al., *Short telomeres on human chromosome 17p. Nat Genet*, 1998. **18**(1): p. 76–80.

52. Mirabello, L., W.Y. Huang, J.Y. Wong, et al., *The association between leukocyte telomere length and cigarette smoking, dietary and physical variables, and risk of prostate cancer. Aging Cell*, 2009. **8**(4): p. 405–13.

53. von Zglinicki, T., G. Saretzki, J. Ladhoff, F. d'Adda di Fagagna, and S.P. Jackson, *Human cell senescence as a DNA damage response. Mech Ageing Dev*, 2005. **126**(1): p. 111–7.

54. Forsyth, N.R., A.P. Evans, J.W. Shay, and W.E. Wright, *Developmental differences in the immortalization of lung fibroblasts by telomerase. Aging Cell*, 2003. **2**(5): p. 235–43.

55. Shiels, P.G., A.J. Kind, K.H. Campbell, et al., *Analysis of telomere lengths in cloned sheep. Nature*, 1999. **399**(6734): p. 316–7.

56. Lanza, R.P., J.B. Cibelli, C. Blackwell, et al., *Extension of cell life-span and telomere length in animals cloned from senescent somatic cells. Science*, 2000. **288**(5466): p. 665–9.

57. Young, N.S., R.T. Calado, and P. Scheinberg, *Current concepts in the pathophysiology and treatment of aplastic anemia. Blood*, 2006. **108**(8): p. 2509–19.

58. Hemann, M.T. and C.W. Greider, *Wild-derived inbred mouse strains have short telomeres. Nucleic Acids Res*, 2000. **28**(22): p. 4474–8.

59. Cawthon, R.M., K.R. Smith, E. O'Brien, A. Sivatchenko, and R.A. Kerber, *Association between telomere length in blood and mortality in people aged 60 years or older. Lancet*, 2003. **361**(9355): p. 393–5.

60. Njajou, O.T., W.C. Hsueh, E.H. Blackburn, et al., *Association between telomere length, specific causes of death, and years of healthy life in health, aging, and body composition, a population-based cohort study. J Gerontol A Biol Sci Med Sci*, 2009. **64**(8): p. 860–4.

61. Bischoff, C., H.C. Petersen, J. Graakjaer, et al., *No association between telomere length and survival among the elderly and oldest old. Epidemiology*, 2006. **17**(2): p. 190–4.

5 The Cellular Senescence Program

Pooja Shivshankar and Claude Jourdan Le Saux

University of Texas Health Science Center, Division of Cardiology and Pulmonary and Critical Care, San Antonio, Texas, USA

5.1 Cellular senescence and evidence of senescence in a cell

Mitosis is the process of somatic cell division that governs development of all organs in the eukaryotic system. Somatic cell division occurs for about 60 doubling times under *in vitro* cultures, called the Hayflick limit, whereby the cells undergo cell cycle arrest [1]. In *in vivo* conditions, the cells also attain the finite level of division, a phenomenon called cellular senescence [2]. Cellular senescence could be classified into two types, irreversible and reversible, depending on the type of response the cells have undergone. Upon series of oxidative stress signals and mitogenic activation, the cells undergo irreversible arrest. These cells remain metabolically active but with altered protein expression profile. Therefore, irreversible cellular senescence is considered as a tumor suppressive mechanism. On the contrary, in reversible type of cellular senescence, oncogenic/mitogenic stimulation and telomere shortening elicit DNA damage response, and post cell cycle arrest, these cells undergo either apoptosis or revert back into transformed phenotypes. Under *in vitro* culture conditions, cellular senescence is classified as the replicative senescence and premature senescence, which correspond to the activation of DNA damage response signals and oxidative stress signals, respectively. In this chapter, we present a broad account of factors that contribute to cellular senescence and how senescent cells in turn contribute to aging and age-associated pathologies of the lungs.

5.1.1 Characteristics of senescent cells and the inflammatory microenvironment

Studies on characteristics of senescence have covered in large using mesenchymal cell cultures under *in vitro* conditions. In recent years, epithelial cell senescence has emerged in age-related tissue damage. Phenotypically, the cells are distended with increased senescence-associated

Molecular Aspects of Aging: Understanding Lung Aging, First Edition. Edited by Mauricio Rojas, Silke Meiners and Claude Jourdan Le Saux.

lysosomal β-galactosidase (SA-β gal) activity being active at low pH. The enzyme activity is encoded by the *glb1* gene on chromosome *3p21.33* [3]. Cellular senescence leads to condensed chromatin structures that control the transcription of select genes and irreversible state of senescence. The phenotypic alteration in senescent cells results in production of excessive growth factors, matrix metalloproteinases, cytoskeletal proteins, and inflammatory mediators, which are collectively known as senescence-associated secretory phenotype (SASP). SASP leads to increased inflammation and angiogenic potential within the microenvironment of the damaged tissue area. The altered activation of signaling pathways increases NF-κB-regulated transcription of genes including production of SASP. Using premature senescent keratinocyte cultures, Bernard D et al. [4] have shown that the apoptotic function of cRel in NF-κB becomes redundant with the upregulation of manganese superoxide dismutase (MnSOD) and survival of the cultures stimulated with tumor necrosis factor alpha (TNFα). The paracrine effect of the SASP subsequently dysregulates normally functioning neighboring cells and promotes malignant transformation and increased susceptibility to microbial assaults. As a tumor suppressor mechanism, senescent cells utilize the SASP for chemotaxis and clearance of tumor cells by phagocytes. The increased innate immune potential of SASP is attributed to the upregulation of toll-like receptor 4, interleukin (IL)-1β, IL-8, monocyte chemotactic protein-1 (MCP-1), intercellular adhesion molecule-1 (ICAM-1), and the chemotactic factor C-X-C motif ligand-1 (CXCL-1). The release of SASP is also associated with the development of age-related diseases such as idiopathic pulmonary fibrosis (IPF). Indeed, TGF-β-induced senescent human bronchoepithelial cells have been implicated in the activation of fibroblasts through the release of IL-1β [5].

Although demonstrated in fibroblasts cultures and the SASP limits to cell-type specificity, similar inflammatory profile was observed in replicative senescence in hepatocytes [6] and the premature senescence model in human type II alveolar epithelial cells induced with genotoxic stress by bleomycin [7, 8]. It has also been demonstrated that oncogenic stress stimulated by *ras* induces premature senescence and generates a replicative senescence phenotype [9]. Furthermore, the proinflammatory cytokine interleukin-1α (IL-1α) primarily transactivates expression of these proinflammatory cytokines and chemokines during SASP generation, which increases NF-κB (binds to IL-8 promoter) and CEBPβ (binds to IL-6 promoter in addition to nuclear factor IL-6 (NFIL-6) activation in the adjacent normal cells. Regulation by IL-1α is mediated through the toll-like receptor pathway coupled with the IL-1 receptor [10]. While senescence is thought to be a potential tumor-suppressing mechanism, owing to its inflammatory microenvironment and paracrine effect, other factors such as the matrix metalloproteinases and growth factors together with the inflammatory mediators tend to promote angiogenesis, tumor progression, and metastasis. Besides cytokines, cyclooxygenase 2, via production of prostaglandins, stimulates angiogenesis and metastatic potential of the tumors [11].

5.1.2 Detection of senescent cells *in vitro* and *in vivo*

Due to the enlarged nuclear size and flattened shape, senescent cells can be morphologically distinguished from their normal counterparts. Initial studies on fibroblast cultures were confined to these apparent changes in the cell size and shape until the lysosomal β-galactosidase activity and reduced telomerase activity were known at cellular level. Aside from these, there have been several markers at molecular level that distinguish senescent phenotypes from the normal cells, such as the tumor suppressor proteins, p53 and the retinoblastoma protein (pRB); the cyclin-dependent kinase inhibitors (Cdkn), Cdkn1 (p21) and Cdkn2 (p16); the chromatin-associated proteins, γH2Ax, in association with activation of checkpoint-kinase 2; and the macrohistone protein 2A (MH2A) complexed with histone 3 chaperone, histone repressor A (HIRA), and chromatin regulator, antisilencing factor 1a (ASF1a) [12, 13, 14]. Recently, a proteomics study on splenic lymphocytes isolated from the senescence-accelerated mouse (SAM) model demonstrated distinct sets of proteins involved in energy metabolism and homoeostasis and macrophage capping and DNA damage as biomarkers of age-associated immune diseases [15].

5.2 Conditions associated with cellular senescence

Cellular senescence has a dual impact in maintaining tissue integrity and homeostasis. This phenomenon has been thought to be a protective process to various stimuli detrimental to the cells such as in cancer cells; the pathways triggered lead to permanent arrest of the cell cycle. Paradoxically, it is thought to be detrimental in aging, as the presence of senescent cells has been associated with increased risk of age-associated diseases, specifically in the lungs, and reduced life span [16, 17, 18, 19]. We have recently demonstrated that senescence might contribute to chronic inflammatory status in aged mouse lungs, as well as young mice that were experimentally induced with oxidative stress and DNA damage responses [20].

5.2.1 Oxidative stress

Stress-induced premature senescence (SIPS) involves oxidative stress as a common characteristic of senescence induction. *In vitro* studies have elucidated the consequences of primary fibroblast cultures isolated from human and murine tissues to undergo senescence via shortened telomeres and p53 activation. In contrast, epithelial cells with feeder layers have shown to adapt the p53 pathway, whereas epithelial cells without feeders take the stress-induced p16 pathway to lead to senescence [21]. These studies nonetheless indicate that the oxidative stress can activate both pathways depending upon the cells environment. This hypothesis is supported by Kurz et al., wherein the endothelial cell senescence by pro-oxidative stress mediators tends to compromise telomerase function under chronic oxidative stressed conditions [22].

On the contrary, mitochondrial response has been shown to increase in oxidative stress-mediated cell cycle arrest. At sublethal concentrations, hydrogen peroxide-treated fibroblasts resulted in increased mitochondrial mass along with permanent cell cycle arrest [23]. The authors further described other cell growth-arresting drugs such as lovastatin, genistein, and mimosine to exhibit similar effects on mitochondrial mass due to oxidative stress. Based on these data, we could hypothesize that oxidative stress and dysregulated mitochondrial function can in turn lead to senescence induction with the notion that age-related lung diseases in humans have increased oxidative stress and lipid peroxidation [24, 25].

5.2.2 DNA damage

Telomere damage-induced senescence (TDIS) has been the most extensively studied mechanism of DNA damage response to mediate senescence during aging. Telomeres are the short repetitive guanine-rich sequence (5′-TTAGGG-3′) at the chromosome ends that are contained and preserved by the telomerase enzyme [26]. All the somatic cells including blood cells undergo telomere shortening upon aging due to spontaneous mutations in the telomerase encoding gene, resulting in impaired telomerase function. The DNA damage response induced with the shortened telomeres triggers the p53/p21 pathway that leads the cells to either enter senescence or induce apoptosis, resulting in organ deterioration. Cells with positive telomere dysfunction-induced foci, also known as TIFs, are recognized as one of the reliable markers of senescence in DNA damage-induced tissue pathologies as well as aging [27, 28].

Although cellular responses to oxidative stress and DNA damage are not quite delineated, it is convincingly known that mitogenic activation, hypoxic conditions that can also cause DNA damage response, results in p53-mediated transactivation of mitochondrial BCL2-associated X protein (Bax), which in turn activates cytochrome C release and caspase-dependent apoptotic pathway [29]. While p53 has been linked to senescence via activation of p21 and nuclear γH2x, decline in apoptotic pathway or reversal to malignant transformation happens with the bypassing of p53-mediated mitochondrial damage during senescence as one possible mechanism. Thus, it may be reasoned that during aging-related lung pathologies, the onset of irreversible senescence mediated by oxidative stress-inducing signals may occur prior to the involvement of p53/p21 due to DNA damage response. Supporting this notion, Baker et al. have demonstrated that mice deficient in inducible p16-positive cells delay age-related disorders of the eyes, muscles, and adipose tissues and, therefore, causally related p16-mediated senescence in age-associated pathologies [30].

5.2.3 Cell cycle arrest and senescence

The cell cycle process involves the four common stages of cell division such as the initial gaps G0 and G1, followed by the synthetic phase, S, that leads to DNA replication. The cells then enter into a small resting phase called G2 followed by the mitotic phase, M, where the cells undergo mitotic division [31]. Normally, the cell cycle arrest occurs at the G0 phase, when the divided cells restart the process of DNA replication. The developed tissues comprise cells at this initial G0 phase with the quiescent phenotype. However, when senescence is stimulated, the cell cycle arrest occurs at both the G0/G1 and the G2/M phases. More importantly, senescent cultures typically arrest at G0/G1 phase, with nuclear condensation resembling the quiescent phenotypes. Later, these presenescent cells enter into the DNA damage response-induced irreversible senescence and arrest at the G2/M phase and have been associated with the formation of senescence-associated heterochromatin foci (SAHF) [32, 33].

5.3 Mechanisms/pathways of senescence induction

The two classic mechanisms of senescence involve the tumor suppressor proteins p53 and p21 and p16 and pRB along with their respective downstream effector proteins as detailed below.

5.3.1 The p53/p21 pathway

DNA damage response induced by telomere attrition and acute genotoxic stress also stimulates the p53/p21 pathway; upon nuclear translocation, p21 interacts with the chromatin-associated protein, γH2x, and leads to the formation of TIFs. Despite the common equivalence between senescence and cell cycle arrest, there exists an actual difference between these two processes. A normal cell cycle arrest can occur at the G1 phase with typical p21 activation. The arrest causes differentiation of the cells and inhibits proliferation. The expression of p21 is generally regulated by p53. However, p21 is also expressed independent of p53, which justifies that mice deficient in p21 do not exhibit susceptibility to tumor development, suggesting that senescence-associated cell cycle arrest can be induced by p21 independent of p53.

5.3.2 The p16/pRB pathway

The p16/pRB pathway mediates cell senescence upon stimulation with oncogenes and oxidative stress. The downstream step involves the interaction of histone 2A variant (macroH2A) with the histone 3 chaperone, HIRA, and chromatin regulator, ASF1a, resulting in the formation of SAHF. Permanent cell cycle arrest is significantly contributed by SAHF by irreversibly shutting expression of proliferation-associated genes. Although the p53/p21 and p16/pRB pathways converge at different levels depending on the types of cumulative exposures, studies have differentiated these pathways on the basis of distinct sets of nuclear proteins that mediate senescence [34]. For example, age-related telomere attrition induces p53, and reactive oxygen species (ROS) generated by acute genotoxic stress may also lead to p53-mediated DNA damage response [35]. Therefore, it is important to highlight how these proteins perform distinct roles individually in different pathways, and the convergence of their function is tightly controlled during the induction of cellular senescence.

5.3.3 Convergence/coactivation of p53/p21 and p16/pRB pathways

Activation of p16 is evident after p53-mediated p21 activation and plays as an independent pathway together with pRB. pRB (in its nonphosphorylated form) is activated and binds to the E2F factor in the transcription complex in the G1 phase and stabilizes the complex in its transcriptionally inactive state. This interaction further differentiates this pathway from the p53/p21/γH2x pathway. To ensure the inactive state of transcription, a pRB homologue, p107, is expressed and binds to whatsoever, free E2F protein, resulting in an irreversible senescence phenotype, and it is this state which develops into the SAHF due in part to the interaction of pRB/E2F to other chromatin-associated proteins, MH2A, HIRA, and ASF1. Furthermore, the irreversible cell cycle arrest that leads to senescence phenotype might be quite different from the arrest by p53/p21 that leads to either cellular differentiation or apoptosis. Finally, since p16 is involved in regulating stem cell proliferation during aging, a direct control on cancer development increases the susceptibility of the tissue by senescent phenotypes and SASP.

5.3.4 Induction of senescence via molecular signaling

Transforming growth factor β (TGF-β) plays a pivotal role as an antigrowth signal in cell proliferation and differentiation stages of many cell types. More importantly, senescence was linked to be induced by TGF-β in transformed cell lines under oxidative-stressed conditions [36, 37]. TGF-β1 has also been shown to mediate repression of telomerase activity in both normal and cancer cells through SMAD3 activation and nuclear translocation *in vitro*, and the inhibition appeared to be independent of p53 or pRb signaling pathways [38, 39]. However, it is yet to be understood that activated SMAD3 may interact with one or more of the nuclear factors under *in vivo* conditions and contribute to these major signaling pathways of senescence during aging. More recently, cellular senescence has been induced by TGF-β in primary human bronchoepithelial cells via increased expression of p21 [5].

Caveolin-1 (Cav1), a membrane scaffolding protein, has been reported to mediate senescence during ROS-induced mitogen-activated protein kinase (MAPK) activation [40]. It is interesting to note that Cav1 utilizes p53/p21 signaling to trigger senescence rather than p16/pRB signaling in the presence of oxidative stress resulting in SIPS [41, 42]. We have recently shown that in the presence of DNA damage response and inflammation raised by bleomycin injury, Cav1-deficient mice showed protection against lung fibrosis through inhibition of epithelial cell senescence with reduced pRb/p16 and MH2A levels [43]. More importantly, other studies have clearly indicated the combined influence of the two predominant triggers, that is, oxidative damage and DNA damage responses, on lung cell senescence modulated by Cav1 [8, 44]. Furthermore, senescent chondrocytes in a typical age-related degenerative osteoarthritis has been shown to have increased levels of Cav1 that plays a pivotal role in converging p53/p21 activation along with the activation of pRb, leading to the damaged chondrocytes and progression of the pathology [45].

Mammalian target of rapamycin (mTOR) is yet another important factor involved in senescence induction that gets activated by phosphoinositol 3-kinase (PI3K)-mediated complex mechanism [46]. The PI3K activation in a Janus-faced position mediates cellular senescence via mTOR activation [47, 48]. Consistent with previous reports on attenuating age-related dysfunction by blocking mTOR activation via PI3K [49, 50], our studies have revealed that mTOR is activated in aged mouse lungs, and these mice are responding more to bleomycin-induced lung injury (unpublished data). It is still not understood if inhibition of mTOR is a possible therapeutic option in the treatment of age-related pulmonary diseases. Interestingly, rapamycin, a potent immunomodulatory drug and inhibitor of mTOR, has shown promise in the amelioration of numerous age-associated diseases. Of interest, rapamycin enhances resistance of aged mice to pneumococcal pneumonia through reduced cellular senescence [51]. Rapamycin also prevents epithelial stem cell senescence by reducing oxidative stress through increased MnSOD, suggesting that inhibition of the mTOR pathway could be beneficial in other pulmonary age-related diseases such as chronic obstructive pulmonary disease (COPD) or IPF [52].

It is worth discussing on a natural mechanism triggered in the system to counter senescence that are composed of a family of proteins called silent information regulators 2 (SIRT2) and are collectively named sirtuins [53]. The role of sirtuins in lung emphysema is extensively studied, and it has been demonstrated that SIRT1 inhibits chronic NF-κB activation, thereby reducing proinflammatory cytokine production [54, 55].

5.4 Cellular senescence in aging and age-related diseases of the lungs

With respect to the global effect of SASP on adjacent normal tissue function, it could be speculated that the occurrence of senescence that increases with age-related oxidative stress conditions, such as cardiovascular disease, osteoarthritis, and diabetes mellitus, could promote lung dysfunction [56, 57, 58, 59, 60, 61, 62]. Vascular endothelial cell senescence has been demonstrated in atherosclerotic lesions in the rabbit carotid artery as a potential trigger to endothelial dysfunction in the pulmonary vasculature [63, 64]. Hyperglycemia also increases vascular aging, leading to endothelial cell senescence regulated by apoptosis signal-regulating kinase 1 (ASK-1). These data are supportive of the notion that apoptosis inhibition is one of the common ways to accelerate cell senescence [65]. Thus, besides normal aging as an age-related parameter, senescent phenotypes contribute to the site-specific tissue pathologies and may be implicated in chronic inflammation and pulmonary distress evident in age-associated lung diseases such as COPD and IPF.

5.4.1 Normal aging

The concept of normal aging, regardless of the chronic but persistent inflammation, determines the life span, whereas age-associated abnormalities account for the health span of a given species. Aging has been associated with impaired organ function and increased susceptibility to injury and development of fibrosis. More specifically, aged lungs present enlargement of airspaces similar to tobacco-related emphysema believed to be caused by increased expression of proteases by senescent cells [54, 66]. Aging mechanisms in normal lungs are not well elucidated although increasing evidence suggests a predominant role for cellular senescence. We have recently determined the role of senescence in normal aging mouse lungs and related to increased p21 levels independent of p53; others have reported the suppression of telomerase activity in both human and rodent studies via TGF-β-dependent signaling mechanism through SMAD3 activation [67, 68, 69]. TGF-β-dependent modulation of cell physiology is triggered at the level of senescence. Reflecting the impact of TGF-β activation, these animals also contained increased levels of newly synthesized collagen although not significantly affected at the level of mechanical lung functions (unpublished data). Additionally, we and others have determined that these aged mice are highly susceptible to bleomycin challenge and show increased lung fibrosis and tissue damage versus the young controls [70]. More importantly, senescence is also a consequence of exhaustion of lung tissue repair during aging with repeated damaging assaults [71, 72]. From these studies, it could be reasoned that senescence and the associated inflammatory environment triggers profibrotic phenotypes, resulting in an impaired tissue repair process during bleomycin injury.

5.4.2 Pneumonia

Besides age-associated comorbidities, chronic inflammation by itself is a major risk factor for community-acquired pneumonia, the leading cause of death among the elderly due to *Streptococcus pneumoniae* [73]. Release of toxic components from *S. pneumoniae* such as pneumolysin, an important pneumococcal toxin; cell-wall polysaccharides; phosphorylcholine (ChoP); and the capsule of the pneumococcus itself causes severe damage to the lung vasculature and disseminates to the heart tissues, resulting in plaque formation on the heart valves [74, 75]. Consistently, we have shown that normal aged mice were significantly more permissive to pneumococcal adhesion through increased host receptors, such as keratin 10, laminin receptor, and platelet-activating receptors [20]. Additionally, chronic inflammatory status primed the aged mouse lungs to increased susceptibility to pneumococcal challenge [76]. We further demonstrated that senescent lung cells contributed to the chronic NF-κB activation that also promotes expression of the host receptors for pneumococcal binding and positively correlated to the increased susceptibility of aged mice to pneumococcal infection [20, 77]. Increased susceptibility of aged mice to pneumococcal infection was also due to poor elicitation of innate immune response and defective toll-like receptor function on the alveolar macrophages [78].

Interestingly, *S. pneumoniae*-infected pyogenic muscular abscesses are commonly seen in arthritic patients [79], which is a chronic inflammatory condition. Taken together, these studies clearly indicate that senescent cells at site-specific levels and SASP at the circulatory level might help microbial dissemination from the lungs to the distal parts of the body, resulting in increased risk of multiple-tissue assaults.

5.4.3 Chronic obstructive pulmonary disease

The main causative factor of COPD is smoking, although complications such as asthma, environmental stress, and genetic alterations post exposure to the stress factors are debatable in predisposing to COPD [80, 81]. Chronic low-grade inflammation with progressive aging, along with inflammation caused by COPD, leads to cardiovascular complications, worsening the clinical outcome and increased mortality in these patients [82, 83, 84]. Histology readouts include emphysematous lesions in the lungs with loss of airway epithelial mesh and destruction of the walls of the alveoli as some of the manifestations during pathologic examination of the lung biopsies [85, 86]. Interestingly, lung tissue biopsies obtained from COPD patients and animals exposed to smoking showed NF-κB-induced inflammation and also had senescent type II pneumocytes [72, 77].

Along the same line, we have demonstrated in the mouse model that a more generalized oxidative stress in mice induced by hydrogen peroxide-supplemented drinking water promotes lung cell senescence with epithelial cell injury, alveolar wall destruction, and emphysematous lesions [20]. Furthermore, we demonstrated that even with severe pathologic destructions of the tissue parenchyma, there was an obvious induction of p16 and pRb expression, thereby supporting the notion that cellular senescence might be an important contributor of oxidative stress-induced tissue damage during smoking and occupational exposures.

5.4.4 Idiopathic pulmonary fibrosis

IPF is a chronic disorder of lungs that affects the elderly. Interestingly, the phenomenon of cellular senescence has been proposed in IPF manifestation [87]. It involves injury to the type II pneumocytes and vascular endothelial cells and coagulation. At the molecular level, greater understanding of the disease has been facilitated, as described by Thannickal and Loyd [88] that epithelial regeneration is curtailed with age, and further relates to telomere shortening, one of the aspects of cell senescence demonstrated by Alder et al. [89]. In a recent study by Minagawa et al. [5], β-gal positive senescent epithelial cells and increased levels of p21 were demonstrated in lung biopsies of IPF patients, and they also established, *in vitro*, that TGF-β plays a pivotal role in inducing lung epithelial cell senescence and that the DNA repair-specific sirtuin (SIRT), SIRT6, inhibited TGF-β-induced senescence. TGF-β is a pleiotropic growth factor involved in airway remodeling and fibrosis and has been shown to be an integral component of the pathologic network of lung diseases such as asthma and IPF [90, 91]. Supporting Minagawa et al.'s study, we have recently demonstrated that Cav1 is involved in epithelial cell senescence in mice with bleomycin-induced pulmonary fibrosis [43]. Cav1 has also been implicated in airway remodeling, as an upstream regulatory factor for TGF-β signaling by sequestering TGF-β receptor function [92], suggesting that, similar to PI3K, Cav1 signaling might be mediated by two different facets of activation mechanisms. For instance, both Cav1 and TGF-β have been shown to induce senescence [17]. Given the fact that Cav1 suppresses TGF-β signaling, the spatial and temporal distribution of TGF-β receptors might determine the fate of Cav1-mediated regulation at the cell surface level.

Telomerase activator-driven rescuing of senescent phenotypes has been known for almost a decade [93, 94, 95]. We studied the role of a small-molecule telomerase activator, GRN510 (Geron Inc.), in protecting mice deficient in telomerase activity from bleomycin-induced lung fibrosis [96]. We therefore propose that a concurrent approach with the restoration of telomerase activity along with anti-inflammatory regimens would prevent the cells from undergoing senescence.

5.5 Conclusion

The physiological changes occurring with age in the lungs are not well determined. However, cellular senescence appears to play a critical role in, as well as promotion of, age-related diseases.

References

1. Hayflick L. A brief history of the mortality and immortality of cultured cells. *Keio J Med* 1998;47(3):174–182.
2. Hayflick L. How and why we age. *Exp Gerontol* 1998;33(7–8):639–653.
3. Takano T, Yamanouchi Y. Assignment of human beta-galactosidase-A gene to 3p21.33 by fluorescence in situ hybridization. *Hum Genet* 1993;92(4):403–404.
4. Bernard D, Gosselin K, Monte D, et al. Involvement of rel/nuclear factor-kappaB transcription factors in keratinocyte senescence. *Cancer Res* 2004;64(2):472–481.
5. Minagawa S, Araya J, Numata T, et al. Accelerated epithelial cell senescence in IPF and the inhibitory role of SIRT6 in TGF-beta-induced senescence of human bronchial epithelial cells. *Am J Physiol Lung Cell Mol Physiol* 2011;300(3):L391–L401.
6. Ye X, Meeker HC, Kozlowski PB, et al. Pathological changes in the liver of a senescence accelerated mouse strain (SAMP8): A mouse model for the study of liver diseases. *Histol Histopathol* 2004;19(4):1141–1151.
7. Aoshiba K, Tsuji T, Nagai A. Bleomycin induces cellular senescence in alveolar epithelial cells. *Eur Respir J* 2003;22(3):436–443.
8. Kasper M, Barth K. Bleomycin and its role in inducing apoptosis and senescence in lung cells – modulating effects of caveolin-1. *Curr Cancer Drug Targets* 2009;9(3):341–353.
9. Hutter E, Unterluggauer H, Uberall F, Schramek H, Jansen-Durr P. Replicative senescence of human fibroblasts: The role of ras-dependent signaling and oxidative stress. *Exp Gerontol* 2002;37(10–11):1165–1174.
10. Orjalo AV, Bhaumik D, Gengler BK, Scott GK, Campisi J. Cell surface-bound IL-1alpha is an upstream regulator of the senescence-associated IL-6/IL-8 cytokine network. *Proc Natl Acad Sci USA* 2009;106(40):17031–17036.
11. Hwang ES. Replicative senescence and senescence-like state induced in cancer-derived cells. *Mech Ageing Dev* 2002;123(12):1681–1694.
12. Gire V, Roux P, Wynford-Thomas D, Brondello JM, Dulic V. DNA damage checkpoint kinase Chk2 triggers replicative senescence. *EMBO J* 2004;23(13):2554–2563.
13. Haferkamp S, Tran SL, Becker TM, Scurr LL, Kefford RF, Rizos H. The relative contributions of the p53 and pRb pathways in oncogene-induced melanocyte senescence. *Aging* 2009;1(6):542–556.
14. Oruetxebarria I, Venturini F, Kekarainen T, et al. P16INK4a is required for hSNF5 chromatin remodeler-induced cellular senescence in malignant rhabdoid tumor cells. *J Biol Chem* 2004;279(5):3807–3816.

15. Luo Y, Li Y, Lin C, et al. Comparative proteome analysis of splenic lymphocytes in senescence-accelerated mice. *Gerontology* 2009;55(5):559–569.
16. Calvi CL, Podowski M, D'Alessio FR, et al. Critical transition in tissue homeostasis accompanies murine lung senescence. *PLoS One* 2011;6(6):e20712.
17. Chilosi M, Doglioni C, Murer B, Poletti V. Epithelial stem cell exhaustion in the pathogenesis of idiopathic pulmonary fibrosis. *Sarcoidosis Vasc Diffuse Lung Dis* 2011;27(1):7–18.
18. Chilosi M, Poletti V, Rossi A. The pathogenesis of COPD and IPF: Distinct horns of the same devil? *Respir Res* 2012;13:3.
19. Jane-Wit D, Chun HJ. Mechanisms of dysfunction in senescent pulmonary endothelium. *J Gerontol A Biol Sci Med Sci* 2012;67(3):236–241.
20. Shivshankar P, Boyd AR, Le Saux CJ, Yeh IT, Orihuela CJ. Cellular senescence increases expression of bacterial ligands in the lungs and is positively correlated with increased susceptibility to pneumococcal pneumonia. *Aging Cell* 2011;10(5):798–806.
21. Ramirez RD, Morales CP, Herbert BS, et al. Putative telomere-independent mechanisms of replicative aging reflect inadequate growth conditions. *Genes Dev* 2001;15(4):398–403.
22. Kurz DJ, Decary S, Hong Y, Trivier E, Akhmedov A, Erusalimsky JD. Chronic oxidative stress compromises telomere integrity and accelerates the onset of senescence in human endothelial cells. *J Cell Sci* 2004;117(Pt 11):2417–2426.
23. Lee HC, Yin PH, Lu CY, Chi CW, Wei YH. Increase of mitochondria and mitochondrial DNA in response to oxidative stress in human cells. *Biochem J* 2000;348 Pt 2:425–432.
24. Fahn HJ, Wang LS, Kao SH, Chang SC, Huang MH, Wei YH. Smoking-associated mitochondrial DNA mutations and lipid peroxidation in human lung tissues. *Am J Respir Cell Mol Biol* 1998;19(6):901–909.
25. Lee HC, Lim ML, Lu CY, et al. Concurrent increase of oxidative DNA damage and lipid peroxidation together with mitochondrial DNA mutation in human lung tissues during aging – smoking enhances oxidative stress on the aged tissues. *Arch Biochem Biophys* 1999;362(2):309–316.
26. Griffith JD, Comeau L, Rosenfield S, et al. Mammalian telomeres end in a large duplex loop. *Cell* 1999;97(4):503–514.
27. Herbig U, Ferreira M, Condel L, Carey D, Sedivy JM. Cellular senescence in aging primates. *Science* 2006;311(5765):1257.
28. Jeyapalan JC, Ferreira M, Sedivy JM, Herbig U. Accumulation of senescent cells in mitotic tissue of aging primates. *Mech Ageing Dev* 2007;128(1):36–44.
29. Rozan LM, El-Deiry WS. P53 downstream target genes and tumor suppression: A classical view in evolution. *Cell Death Differ* 2007;14(1):3–9.
30. Baker DJ, Wijshake T, Tchkonia T, et al. Clearance of p16ink4a-positive senescent cells delays ageing-associated disorders. *Nature* 2011;479(7372):232–236.
31. Mathon NF, Lloyd AC. Cell senescence and cancer. *Nat Rev Cancer* 2001;1(3):203–213.
32. Prieur A, Besnard E, Babled A, Lemaitre JM. P53 and p16(ink4a) independent induction of senescence by chromatin-dependent alteration of s-phase progression. *Nat Commun* 2011;2:473.
33. Takahashi A, Ohtani N, Hara E. Irreversibility of cellular senescence: Dual roles of p16ink4a/rb-pathway in cell cycle control. *Cell Div* 2007;2:10.
34. Cruickshanks HA, Adams PD. Chromatin: A molecular interface between cancer and aging. *Curr Opin Genet Dev* 2010;21(1):100–106.
35. Ye X, Zerlanko B, Zhang R, et al. Definition of prb- and p53-dependent and -independent steps in HIRA/ASF1a-mediated formation of senescence-associated heterochromatin foci. *Mol Cell Biol* 2007;27(7):2452–2465.
36. Wu S, Hultquist A, Hydbring P, Cetinkaya C, Oberg F, Larsson LG. Tgf-beta enforces senescence in myc-transformed hematopoietic tumor cells through induction of mad1 and repression of myc activity. *Exp Cell Res* 2009;315(18):3099–3111.
37. Frippiat C, Chen QM, Zdanov S, Magalhaes JP, Remacle J, Toussaint O. Subcytotoxic H_2O_2 stress triggers a release of transforming growth factor-beta 1, which induces biomarkers of cellular senescence of human diploid fibroblasts. *J Biol Chem* 2001;276(4):2531–2537.
38. Katakura Y, Nakata E, Miura T, Shirahata S. Transforming growth factor beta triggers two independent-senescence programs in cancer cells. *Biochem Biophys Res Commun* 1999;255(1):110–115.
39. Kordon EC, McKnight RA, Jhappan C, Hennighausen L, Merlino G, Smith GH. Ectopic TGF beta 1 expression in the secretory mammary epithelium induces early senescence of the epithelial stem cell population. *Dev Biol* 1995;168(1):47–61.

40. Dasari A, Bartholomew JN, Volonte D, Galbiati F. Oxidative stress induces premature senescence by stimulating caveolin-1 gene transcription through p38 mitogen-activated protein kinase/sp1-mediated activation of two GC-rich promoter elements. *Cancer Res* 2006;66(22):10805–10814.

41. Chretien A, Piront N, Delaive E, Demazy C, Ninane N, Toussaint O. Increased abundance of cytoplasmic and nuclear caveolin 1 in human diploid fibroblasts in H(2)O(2)-induced premature senescence and interplay with p38alpha(MAPK). *FEBS Lett* 2008;582(12):1685–1692.

42. Zou H, Stoppani E, Volonte D, Galbiati F. Caveolin-1, cellular senescence and age-related diseases. *Mech Ageing Dev* 2011;132(11–12):533–542.

43. Shivshankar P, Brampton C, Miyasato S, Kasper M, Thannickal VJ, Le Saux CJ. Caveolin-1 deficiency protects from pulmonary fibrosis by modulating epithelial cell senescence in mice. *Am J Respir Cell Mol Biol* 2012;47(1):28–36.

44. Volonte D, Kahkonen B, Shapiro S, Di Y, Galbiati F. Caveolin-1 expression is required for the development of pulmonary emphysema through activation of the ATM–p53–p21 pathway. *J Biol Chem* 2009;284(9):5462–5466.

45. Dai SM, Shan ZZ, Nakamura H, et al. Catabolic stress induces features of chondrocyte senescence through overexpression of caveolin 1: Possible involvement of caveolin 1-induced down-regulation of articular chondrocytes in the pathogenesis of osteoarthritis. *Arthritis Rheum* 2006;54(3):818–831.

46. Bahar R, Hartmann CH, Rodriguez KA, et al. Increased cell-to-cell variation in gene expression in ageing mouse heart. *Nature* 2006;441(7096):1011–1014.

47. Hanaoka N, Tanaka F, Otake Y, et al. Primary lung carcinoma arising from emphysematous bullae. *Lung Cancer* 2002;38(2):185–191.

48. Courtois-Cox S, Genther Williams SM, Reczek EE, et al. A negative feedback signaling network underlies oncogene-induced senescence. *Cancer Cell* 2006;10(6):459–472.

49. Corradetti MN, Guan KL. Upstream of the mammalian target of rapamycin: Do all roads pass through mTOR? *Oncogene* 2006;25(48):6347–6360.

50. Blagosklonny MV. Aging and immortality: Quasi-programmed senescence and its pharmacologic inhibition. *Cell Cycle* 2006;5(18):2087–2102.

51. Hinojosa CA, Mgbemena V, Van Roekel S, et al. Enteric-delivered rapamycin enhances resistance of aged mice to pneumococcal pneumonia through reduced cellular senescence. *Exp Gerontol* 2012;47(12):958–965.

52. Finkel T. Relief with rapamycin: mTOR inhibition protects against radiation-induced mucositis. *Cell Stem Cell* 2012;11(3):287–288.

53. Guarente L. Franklin H. Epstein lecture: Sirtuins, aging, and medicine. *N Engl J Med* 2011; 364(23):2235–2244.

54. Ito K, Barnes PJ. COPD as a disease of accelerated lung aging. *Chest* 2009;135(1):173–180.

55. Yang SR, Wright J, Bauter M, Seweryniak K, Kode A, Rahman I. Sirtuin regulates cigarette smoke-induced proinflammatory mediator release via RelA/p65 NF-kappaB in macrophages in vitro and in rat lungs in vivo: Implications for chronic inflammation and aging. *Am J Physiol Lung Cell Mol Physiol* 2007;292(2):L567–576.

56. Gorgoulis VG, Pratsinis H, Zacharatos P, et al. p53-dependent ICAM-1 overexpression in senescent human cells identified in atherosclerotic lesions. *Lab Invest* 2005;85(4):502–511.

57. Hayashi T, Matsui-Hirai H, Miyazaki-Akita A, et al. Endothelial cellular senescence is inhibited by nitric oxide: Implications in atherosclerosis associated with menopause and diabetes. *Proc Natl Acad Sci USA* 2006;103(45):17018–17023.

58. Martin JA, Brown TD, Heiner AD, Buckwalter JA. Chondrocyte senescence, joint loading and osteoarthritis. *Clin Orthop Relat Res* 2004(427 Suppl):S96–S103.

59. Martin JA, Buckwalter JA. The role of chondrocyte senescence in the pathogenesis of osteoarthritis and in limiting cartilage repair. *J Bone Joint Surg Am* 2003;85-A Suppl 2:106–110.

60. Sone H, Kagawa Y. Pancreatic beta cell senescence contributes to the pathogenesis of type 2 diabetes in high-fat diet-induced diabetic mice. *Diabetologia* 2005;48(1):58–67.

61. Trougakos IP, Poulakou M, Stathatos M, Chalikia A, Melidonis A, Gonos ES. Serum levels of the senescence biomarker clusterin/apolipoprotein J increase significantly in diabetes type II and during development of coronary heart disease or at myocardial infarction. *Exp Gerontol* 2002;37(10–11):1175–1187.

62. Yudoh K, Nguyen T, Nakamura H, Hongo-Masuko K, Kato T, Nishioka K. Potential involvement of oxidative stress in cartilage senescence and development of osteoarthritis: Oxidative stress induces chondrocyte telomere instability and downregulation of chondrocyte function. *Arthritis Res Ther* 2005;7(2):R380–R391.

63. Erusalimsky JD, Kurz DJ. Cellular senescence in vivo: Its relevance in ageing and cardiovascular disease. *Exp Gerontol* 2005;40(8–9):634–642.

64. Folkmann JK, Loft S, Moller P. Oxidatively damaged DNA in aging dyslipidemic ApoE$^{-/-}$ and wild-type mice. *Mutagenesis* 2007;22(2):105–110.

65. Yokoi T, Fukuo K, Yasuda O, et al. Apoptosis signal-regulating kinase 1 mediates cellular senescence induced by high glucose in endothelial cells. *Diabetes* 2006;55(6):1660–1665.

66. Tsuji T, Aoshiba K, Nagai A. Alveolar cell senescence in patients with pulmonary emphysema. *Am J Respir Crit Care Med* 2006;174(8):886–893.

67. Cassar L, Li H, Jiang FX, Liu JP. TGF-beta induces telomerase-dependent pancreatic tumor cell cycle arrest. *Mol Cell Endocrinol* 2010;320(1–2):97–105.

68. Li H, Liu JP. Mechanisms of action of TGF-beta in cancer: Evidence for Smad3 as a repressor of the hTERT gene. *Ann N Y Acad Sci* 2007;1114:56–68.

69. Li H, Xu D, Li J, Berndt MC, Liu JP. Transforming growth factor beta suppresses human telomerase reverse transcriptase (hTERT) by Smad3 interactions with c-Myc and the hTERT gene. *J Biol Chem* 2006;281(35):25588–25600.

70. Sueblinvong V, Neujahr DC, Mills ST, et al. Predisposition for disrepair in the aged lung. *Am J Med Sci* 2012;344(1):41–51.

71. Kim CO, Huh AJ, Han SH, Kim JM. Analysis of cellular senescence induced by lipopolysaccharide in pulmonary alveolar epithelial cells. *Arch Gerontol Geriatr* 2011;54(2):e35–e41.

72. Zhou F, Onizawa S, Nagai A, Aoshiba K. Epithelial cell senescence impairs repair process and exacerbates inflammation after airway injury. *Respir Res* 2011;12:78.

73. Yende S, Tuomanen EI, Wunderink R, et al. Preinfection systemic inflammatory markers and risk of hospitalization due to pneumonia. *Am J Respir Crit Care Med* 2005;172(11):1440–1446.

74. Fillon S, Soulis K, Rajasekaran S, et al. Platelet-activating factor receptor and innate immunity: Uptake of gram-positive bacterial cell wall into host cells and cell-specific pathophysiology. *J Immunol* 2006;177(9):6182–6191.

75. Maerz S, Liu CH, Guo W, Zhu YZ. Anti-ischemic effects of bilobalide on neonatal rat cardiomyocytes and the involvement of the platelet-activating factor receptor. *Biosci Rep* 2011;31(5):439–447.

76. Hinojosa E, Boyd AR, Orihuela CJ. Age-associated inflammation and toll-like receptor dysfunction prime the lungs for pneumococcal pneumonia. *J Infect Dis* 2009;200(4):546–554.

77. Tsuji T, Aoshiba K, Nagai A. Alveolar cell senescence exacerbates pulmonary inflammation in patients with chronic obstructive pulmonary disease. *Respiration* 2010;80(1):59–70.

78. Boyd AR, Shivshankar P, Jiang S, Berton MT, Orihuela CJ. Age-related defects in TLR2 signaling diminish the cytokine response by alveolar macrophages during murine pneumococcal pneumonia. *Exp Gerontol* 2012;47(7):507–518.

79. Garcia-Vidal C, Rodriguez-Fernandez S, Teijon S, et al. Risk factors for opportunistic infections in infliximab-treated patients: The importance of screening in prevention. *Eur J Clin Microbiol Infect Dis* 2009;28(4):331–337.

80. Silva GE, Sherrill DL, Guerra S, Barbee RA. Asthma as a risk factor for COPD in a longitudinal study. *Chest* 2004;126(1):59–65.

81. Zeki AA, Schivo M, Chan A, Albertson TE, Louie S. The asthma-COPD overlap syndrome: A common clinical problem in the elderly. *J Allergy* 2011;2011:861926.

82. Agusti A. Systemic effects of chronic obstructive pulmonary disease: What we know and what we don't know (but should). *Proc Am Thorac Soc* 2007;4(7):522–525.

83. Mentz RJ, Fiuzat M, Wojdyla DM, et al. Clinical characteristics and outcomes of hospitalized heart failure patients with systolic dysfunction and chronic obstructive pulmonary disease: Findings from OPTIMIZE-HF. *Eur J Heart Fail* 2012;14(4):395–403.

84. Barnes PJ, Celli BR. Systemic manifestations and comorbidities of COPD. *Eur Respir J* 2009;33(5):1165–1185.

85. Behzad AR, McDonough JE, Seyednejad N, Hogg JC, Walker DC. The disruption of the epithelial mesenchymal trophic unit in COPD. *COPD* 2009;6(6):421–431.

86. Hogg JC, Timens W. The pathology of chronic obstructive pulmonary disease. *Annu Rev Pathol* 2009;4:435–459.

87. Verma S, Slutsky AS. Idiopathic pulmonary fibrosis – new insights. *N Engl J Med* 2007;356(13):1370–1372.

88. Thannickal VJ, Loyd JE. Idiopathic pulmonary fibrosis: A disorder of lung regeneration? *Am J Respir Crit Care Med* 2008;178(7):663–665.

89. Alder JK, Chen JJ, Lancaster L, et al. Short telomeres are a risk factor for idiopathic pulmonary fibrosis. *Proc Natl Acad Sci USA* 2008;105(35):13051–13056.
90. Kitamura H, Cambier S, Somanath S, et al. Mouse and human lung fibroblasts regulate dendritic cell trafficking, airway inflammation, and fibrosis through integrin αvβ8-mediated activation of TGF-β. *J Clin Invest* 2011;121(7):2863–2875.
91. Makinde T, Murphy RF, Agrawal DK. The regulatory role of TGF-beta in airway remodeling in asthma. *Immunol Cell Biol* 2007;85(5):348–356.
92. Le Saux CJ, Teeters K, Miyasato SK, et al. Down-regulation of caveolin-1, an inhibitor of transforming growth factor-beta signaling, in acute allergen-induced airway remodeling. *J Biol Chem* 2008;283(9):5760–5768.
93. Kunieda T, Minamino T, Katsuno T, et al. Cellular senescence impairs circadian expression of clock genes in vitro and in vivo. *Circ Res* 2006;98(4):532–539.
94. Satyanarayana A, Manns MP, Rudolph KL. Telomeres and telomerase: A dual role in hepatocarcinogenesis. *Hepatology* 2004;40(2):276–283.
95. Tollefsbol TO, Andrews LG. Mechanisms for telomerase gene control in aging cells and tumorigenesis. *Med Hypotheses* 2001;56(6):630–637.
96. Le Saux CJ, Davy P, Brampton C, et al. A novel telomerase activator suppresses lung damage in a murine model of idiopathic pulmonary fibrosis. *PLoS One* 2013;8(3):e58423.

6 Signaling Networks Controlling Cellular Senescence

Leena P. Desai, Yan Y. Sanders, and Victor J. Thannickal

Division of Pulmonary, Allergy, and Critical Care Medicine, Department of Medicine, University of Alabama Birmingham, Birmingham, Alabama, USA

6.1 Introduction

Cellular senescence is classically defined as a state of irreversible growth arrest. This growth arrest typically occurs in the G1 phase of the cell cycle. Senescent cells are characterized by expression of senescence-associated heterochromatin foci (SAHF) and of senescence-associated β-galactosidase (SA-β-gal) activity [1, 2]. Nearly 40 years ago, Hayflick and colleagues [3] reported that primary human diploid fibroblasts (HDFs) have a finite lifespan when cultured *in vitro* and that cultivation over many generations eventually results in a loss of proliferative potential. Thus, the first description of cellular senescence was that of replicative senescence (**intrinsic** senescence) which is mediated by telomere shortening [4, 5, 6]. It has been demonstrated that ectopic expression of telomerases abrogates telomere shortening and replicative senescence of human cells [7].

In addition to replicative senescence, multiple other pathways may lead to cellular senescence. Senescence can be induced by various types of exogenous stressors such as aberrant oncogenic signaling, oxidative stress, and DNA damage. This type of senescence, irrespective of telomere length, may be classified as **extrinsic** or premature senescence. Normal cells can undergo telomere-independent senescence in response to oncogenes, such as activated Ras or Raf [8, 9], and mitogenic signals, such as overexpressed mitogen-activated protein kinase (MAPK) [10] or the E2F-1 transcription factor [11]. Both extrinsic (premature) senescence and intrinsic (replicative) senescence converge on common signaling pathways that give rise to the typical morphological and biochemical features of cellular senescence.

Senescence has been proposed as a tumor suppressor mechanism to prevent emergence of transformed cells [12]. Tumor suppressors that induce cellular senescence include the p16 cyclin-dependent kinase inhibitor (CDKI) [13], the p14/ARF regulator of MDM2 [11], and the promyelocytic leukemia protein (PML) [14]. p16 inhibits kinase activity of Cdk/4,6-cyclin D complexes, which inactivate three retinoblastoma (Rb) family proteins: pRb, p107, and p130 [15]. p53 controls cellular senescence by two major

Molecular Aspects of Aging: Understanding Lung Aging, First Edition. Edited by Mauricio Rojas, Silke Meiners and Claude Jourdan Le Saux.
© 2014 John Wiley & Sons, Inc. Published 2014 by John Wiley & Sons, Inc.

pathways [16, 17]; one is the DNA damage response (DDR) pathway and the other acts through p19^{ARF} protein (p14^{ARF} in human). The INK4a locus and its two products (p19 and p16) are, thus, the central mediators of senescence, upstream to p53 and Rb. The accumulation of p53 leads to the activation of downstream genes such as p21/CIP1 and the induction of cellular senescence or apoptosis depending on cellular context. The p21 protein is an inhibitor of cyclin E/Cdk2 complexes and mediates the activation of Rb by p53. The suppression of wild-type p53 can rescue senescent cells from their senescent phenotype, whereas immortalized cells can be forced into senescence by overexpressing Ras or p16 [18].

This chapter summarizes the molecular signaling pathways that regulate cell senescence, focusing on transcriptional and mechanistic regulation of these pathways in aging. We will discuss how the core signaling pathways, involving the p16 and p53 tumor suppressor genes, are perturbed in cellular senescence (Figure 6.1). Finally, therapeutic approaches to target senescence pathways for the treatment of aging and aging-related diseases are proposed.

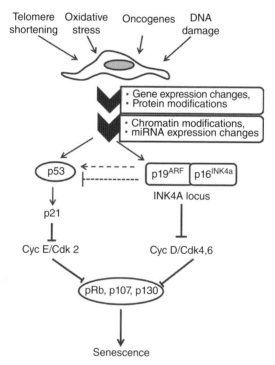

Figure 6.1 Signaling networks regulating cellular senescence. Senescence programs may be triggered by intrinsic or extrinsic mechanisms that converge on downstream signaling cascades that mediate stable cellular growth arrest. ARF, protein that is transcribed from an alternate reading frame of the INK4a/ARF locus; Cyc E, cyclin E; Cdk 2, cyclin-dependent kinase 2; Cyc D: cyclin D; Cdk4,6, cyclin-dependent kinase 4 and 6; miRNA, microRNA; p16-INK4A, a cell cycle regulatory protein that is encoded by the CDKN2a gene; p19-ARF (mouse) or the human equivalent p14-ARF, alternative products of the CDKN2a gene; pRb, retinoblastoma protein.

6.2 Classification of cellular senescence

6.2.1 Intrinsic pathway

Cellular senescence can be broadly classified into the intrinsic (replicative) and extrinsic (premature) pathways. Normal human somatic cells have a finite replicative lifespan, resulting from permanent cell cycle arrest caused by telomere shortening and persistent activation of the DDR [19]. The maintenance of telomere length is essential for DNA replication and cell proliferation, while shortened telomeres can cause persistent activation of the DDR [20]. Experimentally, the relationship between telomere dysfunction and replicative senescence has been investigated by using dominant-negative telomeric repeat factor (TRF2) proteins; in normal human fibroblasts, induction of senescence is attributed to the collapse of the telomere loop which exposes telomeric DNA ends [21, 22]. Thus, it appears that telomere dysfunction is a trigger for induction of replicative senescence. Recently, it has been demonstrated that both age and pathological states are associated with a reduction in cardiac stem cells telomerase activity, telomere shortening, and increased expression of telomere-induced dysfunction, SAHF, p16^{INK4A}, and p21^{CIP} [23]. The molecular mechanisms inducing senescence-like growth arrest in stem cells indicate the involvement of p16^{INK4a} in this process [24]. The induction of p16^{INK4a} inhibits CDKs and downregulates the phosphorylation status of Rb protein, leading to growth arrest [9, 25]. shRNA targeting of p16^{INK4a} [24] or treatment with fibroblast growth factor-2 [26] delays growth arrest in human mesenchymal stem cells (hMSCs). hMSCs can be immortalized by introducing human telomerase reverse transcriptase (hTERT) and Bmi-1, the polycomb protein that suppresses p16^{INK4a} expression [27].

6.2.2 Extrinsic pathway

Cells can be driven to enter a state of senescence without continuous cell division and independent of telomere shortening. Normal cells can be induced to enter senescence by the overexpression of oncogenic Ras or Raf genes [8, 9]. This "oncogene-induced premature senescence" has led to the hypothesis that senescence might have developed as a cellular mechanism to suppress tumor development [28]. In addition, cells can undergo senescence after exposure to a DNA-damaging insult [29]. Such **stress-induced senescence** can be induced in normal as well as cancer cells. Premature senescence is also induced upon chronic exposure to oligomycin, an inhibitor of oxidative phosphorylation, irrespective of reactive oxygen species (ROS) production and which results in the upregulation of p16, p21, and p27 [30]. Thus, there are number of extrinsic inducers of cell senescence that have been demonstrated in different cell types [31].

Senescent cells secrete increased amounts of proinflammatory cytokines and growth factors, both as a cause and consequence of increased cellular stress [32]. Recent studies demonstrate that activated chemokine receptor CXCR2 [33], insulin-like growth factor (IGF)-binding protein 7 [34], IL-6 receptor [35], or downregulation of the transcriptional repressor HES1 [36] may be required for the establishment and/or maintenance of the senescent phenotype in various cell types. In human dermal fibroblasts and endothelial cells, the IGF-binding proteins (IGFBPs) are important members of the IGF signaling pathway that regulate cell senescence via the p53-dependent pathway [37]. IL-1 receptor signaling initiates both miR-146a/b upregulation and cytokine secretion. miR-146a/b is expressed in response to inflammatory cytokines as a negative feedback loop that restrains the senescence-associated secretory phenotype (SASP) [38]. miR-146a downregulates multiple genes associated with inflammation, including IRAK1, IL-6, IL-8, and PAI-1, and inhibits SA-β-gal activity and production of intracellular ROS while increasing cell proliferation [39]. Together, these studies suggest that cellular senescence may represent a continuum, from an initiation stage to a maintenance phase before acquiring a more stable, irreversible, and phenotypically complete senescence.

6.2.3 Reversibility of cellular senescence

It is becoming increasingly clear that there may be some unexpected plasticity to the senescence phenotype. If strictly defined as an **irreversible** process, then the fates of senescent cells are more restricted. With this stricter definition, reversible cell cycle arrest (quiescence) should be distinguished from classical cellular senescence. In some contexts, however, features of senescence such as the characteristic morphology (large, flattened cells), cell cycle arrest, and expression of SA-β-gal may be reversible. Senescence reversal can occur if cells senesce without signaling through the p16–pRb pathway [40]. It is not known whether this occurs *in vivo*. Histologically normal human mammary epithelia have silenced (methylated) p16 promoters [41]; if such cells senesce and subsequently lose p53 function, they could resume proliferation. Senescent p16-positive cells in dysplastic nevi have a mutation that inactivates p53 function and can revert to a proliferating state [42]. P53 is known to inhibit the mammalian target of rapamycin (mTOR) pathway; the inhibition of mTOR by rapamycin converts senescence into quiescence [43, 44]. Thus, p53 is implicated in inducing cellular apoptosis, reversible cell cycle arrest, or senescence.

The recently developed MDM2 antagonists, the nutlins, are effective p53 activators and induce reversible senescence; the retinoblastoma family members (pRb, p107, and p130) are downregulated in cells undergoing nutlin-induced senescence, suggesting a potential mechanism for this reversibility [45]. The expression profiling of human miRs and mRNAs in models of reversible and irreversible senescence has provided additional insights into the mechanisms of reversibility. miRs were found to repress pathways controlling cell cycle, proliferation, DNA repair, and apoptosis, all of which are inhibited in senescent human fibroblasts [46].

6.3 Cross talk of signaling pathways

6.3.1 Protein kinases

Cellular signaling and phenotypes are controlled by key signaling molecules such as cAMP-dependent protein kinase A (PKA) and protein kinase C (PKC). AKAP12/SSeCKS/Gravin (AKAP12) is a scaffold protein for PKA and PKC; AKAP12 deficiency induces hyperactivation of PKC isozymes, leading to Rb-dependent senescence involving PKCα and PKCδ, but not PKCε [47]. Thus, AKAP12 controls MEF senescence through direct scaffolding of PKC isozymes; the expression of full-length AKAP12, but not AKAP12 deleted of its PKC-binding domains, suppresses senescence [47, 48].

PKC (CKII) activity decreases in senescent human lung fibroblast cells and aged rat tissues; CKII inhibition induces premature senescence of human fibroblasts [49] and hMSCs [50].

Antisense inhibitors of four miRNAs (miR-186, miR-216b, miR-337-3p, and miR-760) suppressed cellular senescence and ROS production induced by CKIIα knockdown, whereas overexpression of CKIIα rescued cellular senescence and ROS production induced by the four miRNA mimics [51].

The observation that suppression of PI3K/Akt signaling can result in cellular senescence has been reported [52, 53]. Hyperactive ERK or mTOR signaling has been shown to lead to cellular senescence in normal and cancerous cell lines [54, 55, 56], whereas p16^{INK4A} promotes MSC senescence via suppression of the ERK pathway [57]. Recent evidence shows that c-Jun N-terminal kinase (JNK) is involved in the process of premature senescence. In lung carcinoma cells and primary human embryonic fibroblasts, loss of JNK activity triggers a Bcl-2/ROS/DDR signaling cascade that ultimately leads to premature senescence [58].

6.3.2 Metabolic pathways

Carbohydrate metabolism changes during cellular senescence. Knockdown of cytosolic malate dehydrogenase in young HDFs and human fibroblast cell lines resulted in increased SA-β-gal activity, population doubling time, and p16^{INK4A} and p21^{CIP1} protein levels, while sirtuin 1 (SIRT1) deacetylase, a regulator of cellular senescence, is decreased [59]. Cellular aging and senescence are known to be influenced by metabolic rate [60]. The lifespan of many species can be extended through caloric restriction (CR) [61]; and changes in carbohydrate metabolism may contribute to cellular senescence. In senescent HDF cells, a deregulation of carbohydrate metabolism characterized by an imbalance of glycolytic enzyme activities and a failure to maintain ATP levels was observed; this leads to an upregulation of adenylate kinase activity and levels of AMP, which induces growth-suppressive signals leading to premature senescence [62].

Recent studies reveal two unique functions of p53: regulation of cellular energy metabolism and antioxidant defense functions. The identification of synthesis of cytochrome c oxidase (SCO2) and TP53-induced glycolysis regulator (TIGAR) as two p53 target genes revealed a unique function of p53 in the regulation of energy metabolism and ATP generation [63, 64]. p53 induces the SCO2 gene that is essential for mitochondrial respiration, whereas TIGAR functions to slow glycolysis. In cells undergoing aerobic glycolysis, loss of p53 results in decreased oxygen consumption and impaired mitochondrial respiration, promoting a switch to high glucose utilization. A unique p53 target gene, glutaminase 2 (GLS2), has been identified as a mediator of p53's role in energy metabolism and antioxidant defense [65].

6.3.3 Mitochondria and reactive oxygen species

ROS regulate both cell signaling and, in excess, induce oxidative damage to DNA, lipids, and proteins. Cell signaling may occur in cellular compartments such as cytosol, mitochondria, and nuclei, leading to reversible redox-dependent modifications of proteins. Mitochondrial integrity is critical for cell survival and longevity [66, 67]. Mitochondrial electron transport chain (mETC) dysfunction-induced ROS is linked to the activation of the ASK1/p38 pathway and induction of its downstream targets, p16^{INK4a} and p19^{Arf} [68]. Thus, elevated and sustained expression of stress-response signaling pathways (p38MAPK, SAPK/JNK), a major physiological characteristic of aged tissues, is attributed to mETC dysfunction and is a key factor that promotes stress-induced senescence.

Recent studies suggest cross talk between mitochondria and telomeres. Progressive telomere loss [69] induces p53 activation in aging animals [22, 70, 71], leading to cellular senescence and apoptosis and ultimately contributing to the pathogenesis of age-related diseases. Telomere dysfunction was reported to repress mitochondrial biogenesis and function through the telomere–p53–PGC (peroxisome proliferator-activated receptor gamma, coactivator 1 alpha and beta) signaling pathway, thereby linking telomeres to mitochondrial regulation [72]. In addition, hTERT has been found in the mitochondria to protect cells against oxidative stress and apoptosis [73, 74].

Recently, a positive feedback loop between DDR and ROS production has been demonstrated to induce an irreversible senescent-like phenotype [75]. Fibroblasts from Alzheimer's disease (AD) patients are chronically exposed to oxidative stress that may trigger senescence due to increased ROS production. Increased expression of p21 with decreased Bax levels in AD compared to non-AD cells make AD cells resistant to apoptosis by external oxidative stress [76]. High levels of exogenous oxygen (hyperoxia) elevate intracellular ROS production that contributes to senescence. For example, human fibroblasts cultured in 40% O_2 generate increased ROS production and senesce more quickly than their normoxic controls [77]. Conversely, reducing oxygen levels (hypoxia) extends replicative lifespan of human fibroblasts [78, 79].

Oxidative stress and ROS are modulated by miRNAs. miR-335 and miR-34a promote senescence by suppressing intracellular pathways involving mitochondrial oxidative enzymes, superoxide dismutase 2 (SOD2), and thioredoxin reductase (Txrnd2) [80]. miR-128a inhibits growth of medulloblastoma cells by targeting the Bmi-1 oncogene and altering the intracellular redox state of the tumor cells while promoting cellular senescence.

6.3.4 Integrin and focal adhesion signaling

Integrins are adhesion receptors that transduce biochemical signals from the ECM. Focal adhesions are the sites where cells attach to the substratum via integrin–ECM binding. Various cytoskeletal proteins and signaling molecules, such as focal adhesion kinase (FAK), paxillin, vinculin, talin, p130^{Cas}, and tensin, accumulate in focal adhesions and are tyrosine phosphorylated following integrin activation. The relationship between focal adhesion proteins, the cellular cytoskeleton, and senescence were investigated in immortalized HDFs; mRNAs and intracellular distribution for vinculin and vimentin were reduced, whereas lamin A increased with senescence [81]. However, senescent skin fibroblasts mainly produce vimentin in contrast to actin and tubulin; paxillin and c-Src decrease, whereas p53 anchors to cytoplasmic vimentin in senescent fibroblasts [82]. In HUVECs, depletion of reversion-inducing cysteine-rich protein with Kazal motifs (RECK) gene induces p21^{CIP1} expression by downregulating the activation of β1-integrin and FAK in an MMP-2-dependent manner [83]. In human epidermoid carcinoma cells (A431), downregulation of α5β1 integrin and suppression of epidermal growth factor receptor (EGFR)-mediated signaling led to G1 proliferation arrest [84]; in the same study, increased activation of p53 and expression of p27 was found to be dependent on decreased active FAK. Integrin-linked kinase (ILK) is an integrin-binding cytoplasmic protein that is implicated in accelerating the process of cellular senescence [85]. In young cells from rat primary cardiac fibroblasts, ILK overexpression induces larger cell shapes, lower proliferation capacity, and higher SA-β-gal activity with increased expression of p53 and p21 protein levels, whereas knockdown of ILK prevents phenotypic changes typical of senescence in aging cells [85].

Senescent morphology correlates with enhanced actin stress fibers and redistribution of focal adhesion plaques in senescent-like cells. Early passage HDFs following hydrogen peroxide (H_2O_2) treatment develop senescent morphology via transient elevation of p53 protein and inhibition of Rb hyperphosphorylation [86]. In the same study, vinculin and paxillin localized to the cell periphery in untreated cells, whereas they were randomly distributed throughout stress-induced senescent cells. Similarly, hydrogen peroxide-inducible clone (Hic-5) was initially identified as a gene induced by transforming growth factor-β1 (TGF-β1) or by H_2O_2 [87]. The expression of Hic-5 mRNA is induced during the *in vitro* senescence of HDFs [87]. Forced overexpression of Hic-5 induces senescence-like morphology with increased expression of p21/WAF1/Cip1/sdi1 and ECM proteins, including fibronectin and collagen [88].

6.3.5 Transforming growth factor-β1

TGF-β1 signaling pathways are involved in diverse human pathologies, including cancer, fibrosis, and autoimmune diseases. Recent observations have indicated an emerging role for TGF-β1 in the regulation of mitochondrial oxidative stress responses characteristic of chronic degenerative diseases and aging. Conversely, energy and metabolic pathways regulate TGF-β1 signaling.

TGF-β1 is also implicated in the genetic and epigenetic events leading to metastatic progression via epithelial–mesenchymal transition (EMT). EMT by TGF-β1-mediated Wnt/β-catenin signaling activation resulted in progressive differentiation and senescence [89]. In telomerase-immortalized isogenic basal-like human mammary epithelial cells (HMECs), the expression of a dominant-negative type II receptor (DNRII) of TGF-β1 abrogated autocrine TGF-β signaling, attenuated p21 expression, and suppressed Ras-induced senescence-like growth arrest [90]. However, finite lifespan HMECs undergoing a p16/Rb- and p53-independent oncogene-induced senescence (OIS) in response to oncogenic Ras require TGF-β signaling [91].

TGF-β1 may regulate miRs associated with senescence and aging. Circulating miR-21, a TGF-β1-induced miR [92], is reported to be higher in cardiovascular patients as compared to centenarian offspring [93, 94]. Recently, an miR that is downregulated by TGF-β1, miR-29, modulates cellular phenotypes during aging [95]. TGF-β1-regulated genes associated with fibrosis are derepressed by miR-29 knockdown in lung fibroblasts [96].

6.3.6 Epigenetic mechanisms

Epigenetics is the study of stable alterations in gene expression that are not associated with changes in the DNA sequence. Epigenetics change across lifespan. Mechanistically these changes may occur through chromatin remodeling which modulates gene expression via the alteration of chromatin structure. Several interacting mechanisms regulate chromatin structure and function, including histone modifications, sirtuins (SIRTs), miRs, and DNA methylation [97].

6.3.6.1 Histone modifications

A number of posttranslational modifications of histones regulate chromatin structure and interactions with other epigenetic mechanisms; the main modifications are acetylation, methylation, and phosphorylation. Histone acetylation generally leads to a more relaxed chromatin structure and correlates with increased gene expression. Histone acetylation is regulated by the opposing activities of two groups of enzymes: the histone acetyltrans-ferases, which transfer acetyl groups to histone tail lysines, and the histone deacetylases (HDACs) [97], which remove acetyl groups.

Histone deacetylase 2 (HDAC2) is a class I HDAC that is involved in cell cycle, senescence, proliferation, differentiation, development, apoptosis, and glucocorticoid function in inhibiting inflammatory response. In response to oxidative stress, HDAC2 has a protective effect against DDR and cellular senescence/premature aging via an epigenetic mechanism. Recently, our laboratory has demonstrated that global and locus-specific histone modifica-tions of chromatin regulate altered Bcl-2/Bax gene expression in senescent HDFs, contribut-ing to its apoptosis-resistant phenotype [98].

HDAC2 has been shown to protect against cellular senescence via regulating prosenes-cent genes (i.e., p21 and p16) [99, 100, 101]. In addition, deletion of prosenescent gene p21 protects against cigarette smoke-induced cellular senescence in lung cells and subsequent emphysema [102]. Hence, reduction of HDAC2 may lead to cellular senescence and pulmo-nary emphysema [103]. HDAC2 interacts with poly(ADP-ribose) polymerase-1 (PARP-1) and reduces its acetylation that is required for NF-κB-dependent transcriptional activity [104]. PARP-1/NF-κB signaling cascade is activated during cellular senescence [104].

Increased proinflammatory gene transcription in senescent cells may be due to abnormal histone acetylation and methylation on these gene promoters. In general, histone acetylation is associated with gene activation. Histone H3K4 and H3K36 trimethylation activates chro-matin, whereas histone H3K9 and H3K27 trimethylation silences gene transcription [105–108]. Cigarette smoke extract treatment increases histone H3K4 and H3K36 methyla-tion but reduces H3K9 and H3K27 methylation in lung epithelial cells. It is proposed that histone deacetylation of H3K9 by HDAC2 is a prerequisite for its methylation [109]. Moreover, histone H3K4 methylation limits the accumulation of histone H3K4 acetylation at the gene promoter [110]. These findings suggest the cross-regulation of histone methylation and acetylation, even at the same residue.

6.3.6.2 Sirtuins

SIRTs are a family of seven mammalian NAD+-dependent deacetylases implicated in important regulatory functions in various cellular compartments such as metabolism, cell division, differentiation, survival, and senescence. Increasing evidences suggest that SIRT1 is a critical regulator of stress responses, replicative senescence, inflammation, metabolism, and aging [97]. In human adipose tissue-derived mesenchymal stem cells (hAT-MSCs), the overexpression of miR-486-5p inhibits SIRT1 deacetylase activity and induces premature senescence, suggesting that reduction of adult stem cell self-renewal may be an important mechanism of aging [111]. However, in cancer cells, CDK6, SIRT1, and Sp1, genes which are direct targets of miR-22, were involved in senescence [112]. miR-34a triggers endothe-lial senescence in part through SIRT1, since forced expression of SIRT1 blocks the ability of miR-34a to induce senescence [113, 114]. In young human umbilical vein endothelial cells, human aortic endothelial cells, and human coronary artery endothelial cells, miR-217 induces a premature senescence-like phenotype, leading to impaired angiogenesis via inhibition of SIRT1, modulation of forkhead box O1 (FoxO1), and endothelial nitric oxide synthase acetylation [115]. In old endothelial cells, inhibition of miR-217 reduces senescence and increases angiogenic activity via an increase in SIRT1.

6.3.6.3 MicroRNAs

miRs are a novel class of small, regulatory, noncoding RNA molecules that inhibit the expression of multiple genes at the posttranscriptional level. miRs play a crucial role in development, differentiation, apoptosis, and metabolism and are involved in the pathogenesis of many human diseases [97].

Several lines of evidence support the epigenetic regulation of miR transcription [116]. Lehmann et al. [117] reported that hypermethylation of the miR genome is strongly correlated with the methylation of known tumor suppressor genes, whereas another report found that expression of a large miR cluster (C19MC) was activated in human cancer cells through the demethylation of a CpG-rich region [118]. Thus, the focus has been limited to DNA methylation of CpG-rich regions. It has been demonstrated that histone acetylation plays an important role in chromatin remodeling and is required for gene activation [119, 120]. Although several studies have suggested this possibility, there is no direct evidence showing the regulation of miRNAs by histone modifications in a mammalian system [120, 121]. The cellular senescence of human umbilical cord blood-derived multipotent stem cells (hUCB-MSCs) caused by inhibition of HDAC activity, modification of histones, and transcription of a group of miRs (miR-23a, miR-26a, and miR-30a) leads to the upregulation of p16, p21, and p27 [122]. These studies demonstrate that the activating and repressive epigenetic marks on histones regulate miR transcription, suggesting cross talk between epigenetic regulatory mechanisms, including histone modification, DNA methylation, and miRs.

Multiple miRs have been implicated in regulating cellular senescence. miR-29 was significantly upregulated in studies investigating miRs in klotho-deficient [klotho(−/−)] mice, a senescence model showing phenotypic similarities to premature aging in humans [123], and in normal elderly mice relative to wild-type littermates and young mice [124]. The overexpression of miR-519a in either WI-38 HDFs or human cervical carcinoma HeLa cells triggered senescence, as measured by monitoring SA-β-gal activity [125]. In normal human diploid IMR-90 fibroblasts, a set of senescence-associated miRNAs induced a senescent phenotype by DNA damage [126]. The loss of miR biogenesis activates a DNA damage checkpoint, upregulates p19^{Arf}-p53 signaling, and induces senescence in primary cells [127]. In normal human fibroblasts, induced p53beta with diminished Delta133p53 and increased levels of miR-34a were associated with replicative senescence, but not OIS [128]. miR-29 and miR-30 target the 3′ untranslated region (3′UTR) of B-Myb mRNA, and the modulation of B-Myb expression is known to influence both induced and replicative senescence via activation of the Rb pathway [129]. However, miR-449a regulates growth and senescence of prostate cancer cells through the suppression of Rb phosphorylation and the knockdown of cyclin D1 [130].

The molecular basis for miR-mediated senescence signaling is not well understood. miR-17 and miR-20a are both necessary and sufficient for conferring resistance to Ras-induced senescence by directly targeting p21 [131]. In HMECs, siRNA-mediated knockdown of p21 rescues from Ras-induced senescence involving miR-106b family members [132]. In many cases, one target gene is regulated by multiple miRs, reflecting the complexity of miR function. It has been reported that four miRs (miR-15b, miR-24, miR-25, and miR-141), decreased during senescence, jointly lower expression of the kinase MKK4 [133]. In colon cancer cells, miR-186, miR-216b, miR-337-3p, and miR-760 induce cellular senescence by stimulating ROS production and upregulating p53 and p21 expression and SA-β-gal activity.

6.3.6.4 *DNA methylation*

In mammalian genomic DNA, a methyl group can be added to the cytosine on the fifth base, primarily on CpG dinucleotides. A large portion of CpGs are found in repetitive sequences; some are in clusters called CpG islands, which are usually found in the 5′ region of about half of all human genes [134]. In normal mammalian cells, methylation typically occurs in repetitive sequences, while CpG islands at the 5′ regions are unmethylated. Methylated CpG islands may be observed in special cases, such as in imprinted genes, X-chromosome inactivation, and tissue-specific gene expression [135]. DNA methylation is mediated by DNA methyltransferases (DNMT), which include DNMT1, 3a, and 3b; while DNMT1 is a methylation maintenance enzyme, DNMT 3a and 3b are *de novo* methyltransferases [136].

DNA methylation has an important role in aging. Global genomic DNA methylation level decreases with age in various tissues in mammals [137], yet gene-specific hypermethylation is observed in aging and cancer [134]. Demethylation of satellite DNA is often seen in an age-dependent manner [138], in association with increased chromosomal rearrangements that may lead to cancer. Hypomethylation of CpGs at promoter regions may induce overexpression of proto-oncogenes [139]. For example, in liver tumors, several proto-oncogenes such as c-fos, c-myc, H-ras [140], and Ki-ras [141] genes are hypomethylated. On the other hand, there are specific genes that are hypermethylated, such as the tumor suppressor gene p16, a mechanism for p16 downregulation and important in esophageal cancer development [142]. These observations indicate the important role of DNA methylation in age-related diseases, such as cancer and pulmonary fibrosis. In summary, during the aging process, DNA methylation is decreased on the global level, while it may be increased in a gene-specific manner.

Maximum lifespan in different species may be extended by CR [143]. In human cells, DNA hypermethylation of the E2F-1-binding site blocks access of E2F-1 (an active transcription factor of $p16^{INK4a}$) to the $p16^{INK4a}$ promoter, resulting in $p16^{INK4a}$ downregulation, which contributes to CR-induced lifespan extension. In response to CR, DNMT1 activity significantly increases, thus correcting the overall decreased methylation associated with aging [144]. There are other environmental factors, including diet, that modify DNA methylation. For example, a diet rich in folate (vitamin B12) is known to promote global DNA methylation, while selenium and green tea polyphenols (epigallocatechin-3-gallate (EGCG)) reduce global DNA methylation [145], although effects on gene-specific DNA methylation may vary [146].

6.4 Conclusion

Senescence is involved in normal physiology and aging, as well as in age-related diseases. Senescence has long been considered a physiological response of normal cells to prevent tumor development *in vivo* and possibly as a self-limiting process to halt unremitting fibroproliferation in response to wound healing [147]. However, the role of cellular senescence in age-related diseases such as cancer and fibrosis requires further study. The activation of the cellular senescence program involves multiple molecular pathways that have yet to be fully elucidated. The p16, p53, and Rb pathways function as central mediators for these various signaling pathways.

Cells in culture are exposed to various stresses that lead to progressive accumulation of p16 in some or activation of p53 in others and ultimately to the concomitant activation of both pathways in many cells (Figure 6.1). Different cell types differ in their sensitivity to stress, and the individual contributions of the p53 and the p16 pathways leading to the activation of a senescent phenotype may also differ. The p16 response is more enhanced in human cells, whereas ARF protein function is more prominent in murine cells.

Cellular senescence characterized by cell cycle arrest and various cell biological changes has relied on senescence markers. Kinetic study enables understanding of how cells become fully senescent and facilitates the screening of methods that intervene with cellular senescence [148]. Different senescence phenotypes mature at different rates with different lag times, suggesting that multiple independent mechanisms control the expression of senescence phenotypes. Further work aimed at understanding the role of cell senescence signaling pathways in response to various stimuli that we have discussed in this book chapter may highlight their potential as future drug targets for senescence.

Recently, it has been suggested that cellular senescence plays at least two important roles in normal physiology. Specifically, senescence suppresses tumorigenesis, and there is evidence for a role in terminating the wound-healing response [149, 150]. However, chronic pulmonary diseases such as emphysema and fibrosis are characterized by the presence of senescent cells within remodeled tissues; elucidating mechanisms by which these cells contribute to disease pathogenesis will be essential to develop rational therapeutics for age-related lung disorders.

References

1. Dimri GP, Lee X, Basile G, et al. A biomarker that identifies senescent human cells in culture and in aging skin in vivo. *Proc Natl Acad Sci U S A* 1995;92:9363–9367.
2. Itahana K, Campisi J, Dimri GP. Methods to detect biomarkers of cellular senescence: The senescence-associated beta-galactosidase assay. *Methods Mol Biol* 2007;371:21–31.
3. Hayflick L, Moorhead PS. The serial cultivation of human diploid cell strains. *Exp Cell Res* 1961;25:585–621.
4. Harley CB, Futcher AB, Greider CW. Telomeres shorten during ageing of human fibroblasts. *Nature* 1990;345:458–460.
5. Holt SE, Shay JW, Wright WE. Refining the telomere-telomerase hypothesis of aging and cancer. *Nat Biotechnol* 1996;14:836–839.
6. Blackburn EH. Switching and signaling at the telomere. *Cell* 2001;106:661–673.
7. Bodnar AG, Ouellette M, Frolkis M, et al. Extension of life-span by introduction of telomerase into normal human cells. *Science* 1998;279:349–352.
8. Zhu J, Woods D, McMahon M, Bishop JM. Senescence of human fibroblasts induced by oncogenic raf. *Genes Dev* 1998;12:2997–3007.
9. Serrano M, Lin AW, McCurrach ME, Beach D, Lowe SW. Oncogenic ras provokes premature cell senescence associated with accumulation of p53 and p16ink4a. *Cell* 1997;88:593–602.
10. Lin AW, Barradas M, Stone JC, van Aelst L, Serrano M, Lowe SW. Premature senescence involving p53 and p16 is activated in response to constitutive mek/mapk mitogenic signaling. *Genes Dev* 1998;12:3008–3019.
11. Dimri GP, Itahana K, Acosta M, Campisi J. Regulation of a senescence checkpoint response by the e2f1 transcription factor and p14(arf) tumor suppressor. *Mol Cell Biol* 2000;20:273–285.
12. Campisi J, Kim SH, Lim CS, Rubio M. Cellular senescence, cancer and aging: The telomere connection. *Exp Gerontol* 2001;36:1619–1637.
13. McConnell BB, Starborg M, Brookes S, Peters G. Inhibitors of cyclin-dependent kinases induce features of replicative senescence in early passage human diploid fibroblasts. *Curr Biol* 1998;8:351–354.
14. Ferbeyre G, de Stanchina E, Querido E, Baptiste N, Prives C, Lowe SW. Pml is induced by oncogenic ras and promotes premature senescence. *Genes Dev* 2000;14:2015–2027.
15. Amati B, Alevizopoulos K, Vlach J. Myc and the cell cycle. *Front Biosci* 1998;3:d250–d268.

16. Wahl GM, Carr AM. The evolution of diverse biological responses to DNA damage: Insights from yeast and p53. *Nat Cell Biol* 2001;3:E277–E286.

17. Sherr CJ. Tumor surveillance via the arf-p53 pathway. *Genes Dev* 1998;12:2984–2991.

18. Harada H, Nakagawa H, Oyama K, et al. Telomerase induces immortalization of human esophageal keratinocytes without p16ink4a inactivation. *Mol Cancer Res* 2003;1:729–738.

19. Ohtani N, Mann DJ, Hara E. Cellular senescence: Its role in tumor suppression and aging. *Cancer Sci* 2009;100:792–797.

20. Sikora E, Arendt T, Bennett M, Narita M. Impact of cellular senescence signature on ageing research. *Ageing Res Rev* 2011;10:146–152.

21. Smogorzewska A, Karlseder J, Holtgreve-Grez H, Jauch A, de Lange T. DNA ligase iv-dependent nhej of deprotected mammalian telomeres in g1 and g2. *Curr Biol* 2002;12:1635–1644.

22. Karlseder J, Broccoli D, Dai Y, Hardy S, de Lange T. P53- and atm-dependent apoptosis induced by telomeres lacking trf2. *Science* 1999;283:1321–1325.

23. Cesselli D, Beltrami AP, D'Aurizio F, et al. Effects of age and heart failure on human cardiac stem cell function. *Am J Pathol* 2011;179:349–366.

24. Shibata KR, Aoyama T, Shima Y, et al. Expression of the p16ink4a gene is associated closely with senescence of human mesenchymal stem cells and is potentially silenced by DNA methylation during in vitro expansion. *Stem Cells* 2007;25:2371–2382.

25. Alcorta DA, Xiong Y, Phelps D, Hannon G, Beach D, Barrett JC. Involvement of the cyclin-dependent kinase inhibitor p16 (ink4a) in replicative senescence of normal human fibroblasts. *Proc Natl Acad Sci U S A* 1996;93:13742–13747.

26. Ito T, Sawada R, Fujiwara Y, Seyama Y, Tsuchiya T. Fgf-2 suppresses cellular senescence of human mesenchymal stem cells by down-regulation of tgf-beta2. *Biochem Biophys Res Commun* 2007;359:108–114.

27. Zhang X, Soda Y, Takahashi K, et al. Successful immortalization of mesenchymal progenitor cells derived from human placenta and the differentiation abilities of immortalized cells. *Biochem Biophys Res Commun* 2006;351:853–859.

28. Campisi J, d'Adda di Fagagna F. Cellular senescence: When bad things happen to good cells. *Nat Rev Mol Cell Biol* 2007;8:729–740.

29. Toussaint O, Medrano EE, von Zglinicki T. Cellular and molecular mechanisms of stress-induced premature senescence (sips) of human diploid fibroblasts and melanocytes. *Exp Gerontol* 2000;35:927–945.

30. Stockl P, Hutter E, Zwerschke W, Jansen-Durr P. Sustained inhibition of oxidative phosphorylation impairs cell proliferation and induces premature senescence in human fibroblasts. *Exp Gerontol* 2006;41:674–682.

31. Beltrami AP, Cesselli D, Beltrami CA. At the stem of youth and health. *Pharmacol Ther* 2011;129:3–20.

32. Korybalska K, Kawka E, Kusch A, et al. Recovery of senescent endothelial cells from injury. *J Gerontol A Biol Sci Med Sci* 2013;68(3):250–257.

33. Acosta JC, O'Loghlen A, Banito A, et al. Chemokine signaling via the cxcr2 receptor reinforces senescence. *Cell* 2008;133:1006–1018.

34. Wajapeyee N, Serra RW, Zhu X, Mahalingam M, Green MR. Oncogenic braf induces senescence and apoptosis through pathways mediated by the secreted protein igfbp7. *Cell* 2008;132:363–374.

35. Kuilman T, Michaloglou C, Vredeveld LC, et al. Oncogene-induced senescence relayed by an interleukin-dependent inflammatory network. *Cell* 2008;133:1019–1031.

36. Sang L, Coller HA, Roberts JM. Control of the reversibility of cellular quiescence by the transcriptional repressor hes1. *Science* 2008;321:1095–1100.

37. Kim KS, Seu YB, Baek SH, et al. Induction of cellular senescence by insulin-like growth factor binding protein-5 through a p53-dependent mechanism. *Mol Biol Cell* 2007;18:4543–4552.

38. Bhaumik D, Scott GK, Schokrpur S, et al. MicroRNA$_s$ mir-146a/b negatively modulate the senescence-associated inflammatory mediators il-6 and il-8. *Aging (Albany NY)* 2009;1:402–411.

39. Li G, Luna C, Qiu J, Epstein DL, Gonzalez P. Modulation of inflammatory markers by mir-146a during replicative senescence in trabecular meshwork cells. *Invest Ophthalmol Vis Sci* 2010;51:2976–2985.

40. Beausejour CM, Krtolica A, Galimi F, et al. Reversal of human cellular senescence: Roles of the p53 and p16 pathways. *EMBO J* 2003;22:4212–4222.

41. Holst CR, Nuovo GJ, Esteller M, et al. Methylation of p16(ink4a) promoters occurs in vivo in histologically normal human mammary epithelia. *Cancer Res* 2003;63:1596–1601.

42. Michaloglou C, Vredeveld LC, Soengas MS, et al. Brafe600-associated senescence-like cell cycle arrest of human naevi. *Nature* 2005;436:720–724.
43. Leontieva OV, Natarajan V, Demidenko ZN, Burdelya LG, Gudkov AV, Blagosklonny MV. Hypoxia suppresses conversion from proliferative arrest to cellular senescence. *Proc Natl Acad Sci U S A* 2012;109:13314–13318.
44. Demidenko ZN, Zubova SG, Bukreeva EI, Pospelov VA, Pospelova TV, Blagosklonny MV. Rapamycin decelerates cellular senescence. *Cell Cycle* 2009;8:1888–1895.
45. Huang B, Deo D, Xia M, Vassilev LT. Pharmacologic p53 activation blocks cell cycle progression but fails to induce senescence in epithelial cancer cells. *Mol Cancer Res* 2009;7:1497–1509.
46. Maes OC, Sarojini H, Wang E. Stepwise up-regulation of microRNA expression levels from replicating to reversible and irreversible growth arrest states in wi-38 human fibroblasts. *J Cell Physiol* 2009;221:109–119.
47. Akakura S, Nochajski P, Gao L, Sotomayor P, Matsui S, Gelman IH. Rb-dependent cellular senescence, multinucleation and susceptibility to oncogenic transformation through pkc scaffolding by ssecks/akap12. *Cell Cycle* 2010;9:4656–4665.
48. Akakura S, Gelman IH. Pivotal role of akap12 in the regulation of cellular adhesion dynamics: Control of cytoskeletal architecture, cell migration, and mitogenic signaling. *J Signal Transduct* 2012;2012:529179.
49. Ryu SW, Woo JH, Kim YH, Lee YS, Park JW, Bae YS. Downregulation of protein kinase ckii is associated with cellular senescence. *FEBS Lett* 2006;580:988–994.
50. Wang D, Jang DJ. Protein kinase ck2 regulates cytoskeletal reorganization during ionizing radiation-induced senescence of human mesenchymal stem cells. *Cancer Res* 2009;69:8200–8207.
51. Kim SY, Lee YH, Bae YS. Mir-186, mir-216b, mir-337-3p, and mir-760 cooperatively induce cellular senescence by targeting alpha subunit of protein kinase ckii in human colorectal cancer cells. *Biochem Biophys Res Commun* 2012;429:173–179.
52. Collado M, Medema RH, Garcia-Cao I, et al. Inhibition of the phosphoinositide 3-kinase pathway induces a senescence-like arrest mediated by p27kip1. *J Biol Chem* 2000;275:21960–21968.
53. Axanova LS, Chen YQ, McCoy T, Sui G, Cramer SD. 1,25-dihydroxyvitamin d(3) and pi3k/akt inhibitors synergistically inhibit growth and induce senescence in prostate cancer cells. *Prostate* 2010;70:1658–1671.
54. Sumikawa E, Matsumoto Y, Sakemura R, Fujii M, Ayusawa D. Prolonged unbalanced growth induces cellular senescence markers linked with mechano transduction in normal and tumor cells. *Biochem Biophys Res Commun* 2005;335:558–565.
55. Korotchkina LG, Leontieva OV, Bukreeva EI, Demidenko ZN, Gudkov AV, Blagosklonny MV. The choice between p53-induced senescence and quiescence is determined in part by the mtor pathway. *Aging (Albany NY)* 2010;2:344–352.
56. Demidenko ZN, Shtutman M, Blagosklonny MV. Pharmacologic inhibition of mek and pi-3k converges on the mtor/s6 pathway to decelerate cellular senescence. *Cell Cycle* 2009;8:1896–1900.
57. Gu Z, Cao X, Jiang J, et al. Upregulation of p16(ink4a) promotes cellular senescence of bone marrow-derived mesenchymal stem cells from systemic lupus erythematosus patients. *Cell Signal* 2012;24:2307–2314.
58. Lee JJ, Lee JH, Ko YG, Hong SI, Lee JS. Prevention of premature senescence requires jnk regulation of bcl-2 and reactive oxygen species. *Oncogene* 2010;29:561–575.
59. Lee SM, Dho SH, Ju SK, Maeng JS, Kim JY, Kwon KS. Cytosolic malate dehydrogenase regulates senescence in human fibroblasts. *Biogerontology* 2012;13:525–536.
60. Jazwinski SM. Metabolic control and ageing. *Trends Genet* 2000;16:506–511.
61. Lane MA, Black A, Handy A, Tilmont EM, Ingram DK, Roth GS. Caloric restriction in primates. *Ann N Y Acad Sci* 2001;928:287–295.
62. Zwerschke W, Mazurek S, Stockl P, Hutter E, Eigenbrodt E, Jansen-Durr P. Metabolic analysis of senescent human fibroblasts reveals a role for amp in cellular senescence. *Biochem J* 2003;376:403–411.
63. Matoba S, Kang JG, Patino WD, et al. P53 regulates mitochondrial respiration. *Science* 2006;312:1650–1653.
64. Bensaad K, Tsuruta A, Selak MA, et al. Tigar, a p53-inducible regulator of glycolysis and apoptosis. *Cell* 2006;126:107–120.
65. Hu W, Zhang C, Wu R, Sun Y, Levine A, Feng Z. Glutaminase 2, a novel p53 target gene regulating energy metabolism and antioxidant function. *Proc Natl Acad Sci U S A* 2010;107:7455–7460.

66. Kirkwood TB. Understanding the odd science of aging. *Cell* 2005;120:437–447.

67. Wallace DC. A mitochondrial paradigm of metabolic and degenerative diseases, aging, and cancer: A dawn for evolutionary medicine. *Annu Rev Genet* 2005;39:359–407.

68. Papaconstantinou J, Hsieh CC. Activation of senescence and aging characteristics by mitochondrially generated ros: How are they linked? *Cell Cycle* 2010;9:3831–3833.

69. Zou Y, Yi X, Wright WE, Shay JW. Human telomerase can immortalize Indian muntjac cells. *Exp Cell Res* 2002;281:63–76.

70. Smogorzewska A, de Lange T. Different telomere damage signaling pathways in human and mouse cells. *EMBO J* 2002;21:4338–4348.

71. van Steensel B, Smogorzewska A, de Lange T. Trf2 protects human telomeres from end-to-end fusions. *Cell* 1998;92:401–413.

72. Sahin E, Colla S, Liesa M, et al. Telomere dysfunction induces metabolic and mitochondrial compromise. *Nature* 2011;470:359–365.

73. Ahmed S, Passos JF, Birket MJ, et al. Telomerase does not counteract telomere shortening but protects mitochondrial function under oxidative stress. *J Cell Sci* 2008;121:1046–1053.

74. Passos JF, Saretzki G, von Zglinicki T. DNA damage in telomeres and mitochondria during cellular senescence: Is there a connection? *Nucleic Acids Res* 2007;35:7505–7513.

75. Passos JF, Nelson G, Wang C, et al. Feedback between p21 and reactive oxygen production is necessary for cell senescence. *Mol Syst Biol* 2010;6:347.

76. Naderi J, Lopez C, Pandey S. Chronically increased oxidative stress in fibroblasts from alzheimer's disease patients causes early senescence and renders resistance to apoptosis by oxidative stress. *Mech Ageing Dev* 2006;127:25–35.

77. Saretzki G, Feng J, von Zglinicki T, Villeponteau B. Similar gene expression pattern in senescent and hyperoxic-treated fibroblasts. *J Gerontol A Biol Sci Med Sci* 1998;53:B438–B442.

78. Packer L, Fuehr K. Low oxygen concentration extends the lifespan of cultured human diploid cells. *Nature* 1977;267:423–425.

79. Saito H, Hammond AT, Moses RE. The effect of low oxygen tension on the in vitro-replicative life span of human diploid fibroblast cells and their transformed derivatives. *Exp Cell Res* 1995;217:272–279.

80. Bai XY, Ma Y, Ding R, Fu B, Shi S, Chen XM. Mir-335 and mir-34a promote renal senescence by suppressing mitochondrial antioxidative enzymes. *J Am Soc Nephrol* 2011;22:1252–1261.

81. Kaneko S, Satoh Y, Ikemura K, et al. Alterations of expression of the cytoskeleton after immortalization of human fibroblasts. *Cell Struct Funct* 1995;20:107–115.

82. Nishio K, Inoue A. Senescence-associated alterations of cytoskeleton: Extraordinary production of vimentin that anchors cytoplasmic p53 in senescent human fibroblasts. *Histochem Cell Biol* 2005;123:263–273.

83. Miki T, Shamma A, Kitajima S, et al. The ss1-integrin-dependent function of reck in physiologic and tumor angiogenesis. *Mol Cancer Res* 2010;8:665–676.

84. Morozevich GE, Kozlova NI, Ushakova NA, Preobrazhenskaya ME, Berman AE. Integrin alpha5beta1 simultaneously controls egfr-dependent proliferation and akt-dependent pro-survival signaling in epidermoid carcinoma cells. *Aging (Albany NY)* 2012;4:368–374.

85. Chen X, Li Z, Feng Z, et al. Integrin-linked kinase induces both senescence-associated alterations and extracellular fibronectin assembly in aging cardiac fibroblasts. *J Gerontol A Biol Sci Med Sci* 2006;61:1232–1245.

86. Chen QM, Tu VC, Catania J, Burton M, Toussaint O, Dilley T. Involvement of rb family proteins, focal adhesion proteins and protein synthesis in senescent morphogenesis induced by hydrogen peroxide. *J Cell Sci* 2000;113 (Pt 22):4087–4097.

87. Shibanuma M, Mashimo J, Kuroki T, Nose K. Characterization of the tgf beta 1-inducible hic-5 gene that encodes a putative novel zinc finger protein and its possible involvement in cellular senescence. *J Biol Chem* 1994;269:26767–26774.

88. Shibanuma M, Mochizuki E, Maniwa R, et al. Induction of senescence-like phenotypes by forced expression of hic-5, which encodes a novel lim motif protein, in immortalized human fibroblasts. *Mol Cell Biol* 1997;17:1224–1235.

89. Kawakita T, Espana EM, Higa K, Kato N, Li W, Tseng SC. Activation of smad-mediated tgf-beta signaling triggers epithelial-mesenchymal transitions in murine cloned corneal progenitor cells. *J Cell Physiol* 2013;228:225–234.

90. Lin S, Yang J, Elkahloun AG, et al. Attenuation of tgf-beta signaling suppresses premature senescence in a p21-dependent manner and promotes oncogenic ras-mediated metastatic transformation in human mammary epithelial cells. *Mol Biol Cell* 2012;23:1569–1581.

91. Cipriano R, Kan CE, Graham J, Danielpour D, Stampfer M, Jackson MW. Tgf-beta signaling engages an atm-chk2-p53-independent ras-induced senescence and prevents malignant transformation in human mammary epithelial cells. *Proc Natl Acad Sci U S A* 2011;108:8668–8673.

92. Yang S, Xie N, Cui H, et al. Mir-31 is a negative regulator of fibrogenesis and pulmonary fibrosis. *FASEB J* 2012;26:3790–3799.

93. Olivieri F, Spazzafumo L, Santini G, et al. Age-related differences in the expression of circulating microRNAs: Mir-21 as a new circulating marker of inflammaging. *Mech Ageing Dev* 2012;133 (11–12):675–685.

94. Olivieri F, Lazzarini R, Recchioni R, et al. Mir-146a as marker of senescence-associated pro-inflammatory status in cells involved in vascular remodelling. *Age (Dordr)* 2013;35(4):1157–1172.

95. Milewicz DM. MicroRNAs, fibrotic remodeling, and aortic aneurysms. *J Clin Invest* 2012;122: 490–493.

96. Cushing L, Kuang PP, Qian J, et al. Mir-29 is a major regulator of genes associated with pulmonary fibrosis. *Am J Respir Cell Mol Biol* 2011;45:287–294.

97. Kimura A, Matsubara K, Horikoshi M. A decade of histone acetylation: Marking eukaryotic chromosomes with specific codes. *J Biochem* 2005;138:647–662.

98. Sanders YY, Liu H, Zhang X, et al. Histone modifications in senescence associated resistance to apoptosis by oxidative stress. *Redox Biol* 2013;1:8–16.

99. Wagner M, Brosch G, Zwerschke W, Seto E, Loidl P, Jansen-Durr P. Histone deacetylases in replicative senescence: Evidence for a senescence-specific form of hdac-2. *FEBS Lett* 2001;499:101–106.

100. Zhou Q, Wang Y, Yang L, et al. Histone deacetylase inhibitors blocked activation and caused senescence of corneal stromal cells. *Mol Vis* 2008;14:2556–2565.

101. Zhou R, Han L, Li G, Tong T. Senescence delay and repression of p16ink4a by lsh via recruitment of histone deacetylases in human diploid fibroblasts. *Nucleic Acids Res* 2009;37:5183–5196.

102. Alder JK, Guo N, Kembou F, et al. Telomere length is a determinant of emphysema susceptibility. *Am J Respir Crit Care Med* 2011;184:904–912.

103. Yao H, Chung S, Hwang JW, et al. Sirt1 protects against emphysema via foxo3-mediated reduction of premature senescence in mice. *J Clin Invest* 2012;122:2032–2045.

104. Hassa PO, Haenni SS, Buerki C, et al. Acetylation of poly(adp-ribose) polymerase-1 by p300/creb-binding protein regulates coactivation of nf-kappab-dependent transcription. *J Biol Chem* 2005;280:40450–40464.

105. De Santa F, Totaro MG, Prosperini E, Notarbartolo S, Testa G, Natoli G. The histone h3 lysine-27 demethylase jmjd3 links inflammation to inhibition of polycomb-mediated gene silencing. *Cell* 2007;130:1083–1094.

106. El Gazzar M, Yoza BK, Hu JY, Cousart SL, McCall CE. Epigenetic silencing of tumor necrosis factor alpha during endotoxin tolerance. *J Biol Chem* 2007;282:26857–26864.

107. Foster SL, Hargreaves DC, Medzhitov R. Gene-specific control of inflammation by tlr-induced chromatin modifications. *Nature* 2007;447:972–978.

108. Nottke A, Colaiacovo MP, Shi Y. Developmental roles of the histone lysine demethylases. *Development* 2009;136:879–889.

109. Nakayama J, Rice JC, Strahl BD, Allis CD, Grewal SI. Role of histone h3 lysine 9 methylation in epigenetic control of heterochromatin assembly. *Science* 2001;292:110–113.

110. Guillemette B, Drogaris P, Lin HH, et al. H3 lysine 4 is acetylated at active gene promoters and is regulated by h3 lysine 4 methylation. *PLoS Genet* 2011;7:e1001354.

111. Kim YJ, Hwang SH, Lee SY, et al. Mir-486-5p induces replicative senescence of human adipose tissue-derived mesenchymal stem cells and its expression is controlled by high glucose. *Stem Cells Dev* 2012;21:1749–1760.

112. Xu D, Takeshita F, Hino Y, et al. Mir-22 represses cancer progression by inducing cellular senescence. *J Cell Biol* 2011;193:409–424.

113. Ito T, Yagi S, Yamakuchi M. MicroRNA-34a regulation of endothelial senescence. *Biochem Biophys Res Commun* 2010;398:735–740.

114. Zhao T, Li J, Chen AF. MicroRNA-34a induces endothelial progenitor cell senescence and impedes its angiogenesis via suppressing silent information regulator 1. *Am J Physiol Endocrinol Metab* 2010;299:E110–E116.

115. Menghini R, Casagrande V, Cardellini M, et al. MicroRNA 217 modulates endothelial cell senescence via silent information regulator 1. *Circulation* 2009;120:1524–1532.

116. Saito Y, Jones PA. Epigenetic activation of tumor suppressor microRNAs in human cancer cells. *Cell Cycle* 2006;5:2220–2222.

117. Lehmann U, Hasemeier B, Christgen M, et al. Epigenetic inactivation of microRNA gene hsa-mir-9-1 in human breast cancer. *J Pathol* 2008;214:17–24.

118. Tsai KW, Kao HW, Chen HC, Chen SJ, Lin WC. Epigenetic control of the expression of a primate-specific microRNA cluster in human cancer cells. *Epigenetics* 2009;4:587–592.

119. Kim W, Benhamed M, Servet C, et al. Histone acetyltransferase gcn5 interferes with the mirna pathway in *Arabidopsis. Cell Res* 2009;19:899–909.

120. Bandres E, Agirre X, Bitarte N, et al. Epigenetic regulation of microRNA expression in colorectal cancer. *Int J Cancer* 2009;125:2737–2743.

121. Scott GK, Mattie MD, Berger CE, Benz SC, Benz CC. Rapid alteration of microRNA levels by histone deacetylase inhibition. *Cancer Res* 2006;66:1277–1281.

122. Lee S, Jung JW, Park SB, et al. Histone deacetylase regulates high mobility group a2-targeting microRNAs in human cord blood-derived multipotent stem cell aging. *Cell Mol Life Sci* 2011;68:325–336.

123. Kuro-o M, Matsumura Y, Aizawa H, et al. Mutation of the mouse klotho gene leads to a syndrome resembling ageing. *Nature* 1997;390:45–51.

124. Takahashi M, Eda A, Fukushima T, Hohjoh H. Reduction of type iv collagen by upregulated mir-29 in normal elderly mouse and klotho-deficient, senescence-model mouse. *PLoS One* 2012;7:e48974.

125. Marasa BS, Srikantan S, Martindale JL, et al. MicroRNA profiling in human diploid fibroblasts uncovers mir-519 role in replicative senescence. *Aging (Albany NY)* 2010;2:333–343.

126. Faraonio R, Salerno P, Passaro F, et al. A set of mirnas participates in the cellular senescence program in human diploid fibroblasts. *Cell Death Differ* 2012;19:713–721.

127. Mudhasani R, Zhu Z, Hutvagner G, et al. Loss of mirna biogenesis induces p19arf-p53 signaling and senescence in primary cells. *J Cell Biol* 2008;181:1055–1063.

128. Fujita K, Mondal AM, Horikawa I, et al. P53 isoforms delta133p53 and p53beta are endogenous regulators of replicative cellular senescence. *Nat Cell Biol* 2009;11:1135–1142.

129. Martinez I, Cazalla D, Almstead LL, Steitz JA, DiMaio D. Mir-29 and mir-30 regulate b-myb expression during cellular senescence. *Proc Natl Acad Sci U S A* 2011;108:522–527.

130. Noonan EJ, Place RF, Basak S, Pookot D, Li LC. Mir-449a causes rb-dependent cell cycle arrest and senescence in prostate cancer cells. *Oncotarget* 2010;1:349–358.

131. Hong L, Lai M, Chen M, et al. The mir-17-92 cluster of microRNAs confers tumorigenicity by inhibiting oncogene-induced senescence. *Cancer Res* 2010;70:8547–8557.

132. Borgdorff V, Lleonart ME, Bishop CL, et al. Multiple microRNAs rescue from ras-induced senescence by inhibiting p21(waf1/cip1). *Oncogene* 2010;29:2262–2271.

133. Marasa BS, Srikantan S, Masuda K, et al. Increased mkk4 abundance with replicative senescence is linked to the joint reduction of multiple microRNAs. *Sci Signal* 2009;2:ra69.

134. Dunn BK. Hypomethylation: One side of a larger picture. *Ann N Y Acad Sci* 2003;983:28–42.

135. Esteller M, Herman JG. Cancer as an epigenetic disease: DNA methylation and chromatin alterations in human tumours. *J Pathol* 2002;196:1–7.

136. Goll MG, Bestor TH. Eukaryotic cytosine methyltransferases. *Annu Rev Biochem* 2005;74:481–514.

137. Fuke C, Shimabukuro M, Petronis A, et al. Age related changes in 5-methylcytosine content in human peripheral leukocytes and placentas: An HPLC-based study. *Ann Hum Genet* 2004;68:196–204.

138. Suzuki T, Fujii M, Ayusawa D. Demethylation of classical satellite 2 and 3 DNA with chromosomal instability in senescent human fibroblasts. *Exp Gerontol* 2002;37:1005–1014.

139. Robertson KD, Jones PA. DNA methylation: Past, present and future directions. *Carcinogenesis* 2000;21:461–467.

140. Rao PM, Antony A, Rajalakshmi S, Sarma DS. Studies on hypomethylation of liver DNA during early stages of chemical carcinogenesis in rat liver. *Carcinogenesis* 1989;10:933–937.

141. Vorce RL, Goodman JI. Altered methylation of ras oncogenes in benzidine-induced b6c3f1 mouse liver tumors. *Toxicol Appl Pharmacol* 1989;100:398–410.

142. Hibi K, Taguchi M, Nakayama H, et al. Molecular detection of p16 promoter methylation in the serum of patients with esophageal squamous cell carcinoma. *Clin Cancer Res* 2001;7:3135–3138.

143. Sinclair DA. Toward a unified theory of caloric restriction and longevity regulation. *Mech Ageing Dev* 2005;126:987–1002.

144. Li Y, Liu L, Tollefsbol TO. Glucose restriction can extend normal cell lifespan and impair precancerous cell growth through epigenetic control of htert and p16 expression. *FASEB J* 2010;24:1442–1453.

145. Davis CD, Uthus EO, Finley JW. Dietary selenium and arsenic affect DNA methylation in vitro in caco-2 cells and in vivo in rat liver and colon. *J Nutr* 2000;130:2903–2909.

146. Zeng H, Yan L, Cheng WH, Uthus EO. Dietary selenomethionine increases exon-specific DNA methylation of the p53 gene in rat liver and colon mucosa. *J Nutr* 2011;141:1464–1468.

147. Krizhanovsky V, Yon M, Dickins RA, et al. Senescence of activated stellate cells limits liver fibrosis. *Cell* 2008;134:657–667.

148. Cho S, Park J, Hwang ES. Kinetics of the cell biological changes occurring in the progression of DNA damage-induced senescence. *Mol Cells* 2011;31:539–546.

149. Rodier F, Campisi J. Four faces of cellular senescence. *J Cell Biol* 2011;192:547–556.

150. Adams PD. Healing and hurting: Molecular mechanisms, functions, and pathologies of cellular senescence. *Mol Cell* 2009;36:2–14.

7 Immune Senescence

Kevin P. High

Section on Infectious Diseases, Wake Forest School of Medicine, Winston-Salem, North Carolina, USA

7.1 Introduction

By 2030, one in five Americans will be age 65 years or older. Though immune responses frequently wane with age, some aspects of immunity (e.g., autoimmunity and inflammation) are enhanced in seniors. Disordered immune responses seen in older adults are collectively referred to as immune senescence which is associated with poor vaccine responses, age-related illnesses (cardiovascular disease, dementia), disability, and death in older adults and even hypothesized to be a *cause* of aging itself.

The immune senescent phenotype is characterized by low-grade inflammation even in the absence of an evident inflammatory stimulus, impaired vaccine responses, and susceptibility to more frequent and severe episodes of infection. There is also evidence that decreased immune **surveillance** may be partially responsible for increased rates of cancer in seniors. Both innate and adaptive immunity are profoundly affected by age, but age itself is only one factor that determines the global status of immunity in a given older individual. Comorbidity (e.g., diabetes, peripheral vascular disease), nutritional status, polypharmacy, physical and cognitive functional impairment, and frequent contact with healthcare and/or institutionalization all markedly alter the host–pathogen relationship, contribute to the immune senescent milieu, and alter colonizing flora. These factors often trump the influence of age itself. The acquisition of comorbid illness occurs throughout a lifetime and is nearly inseparable from age without very carefully controlled studies. Thus, interpretation of concepts in the literature must be tempered by the understanding that comorbidity is a frequent and robust confounder but often unmeasured and/or unreported in many studies. An important corollary, however, is that attempts to isolate the effect of age when studying immunity (e.g., use of very strict criteria such as SENIEUR) may scientifically enhance the validity of attributing a change in immunity to age itself but increasingly limits the clinical generalizability/importance of noted changes. Thus, it is critically important to examine the study design and population inclusion/exclusion criteria when considering the findings of any given study.

Molecular Aspects of Aging: Understanding Lung Aging, First Edition. Edited by Mauricio Rojas, Silke Meiners and Claude Jourdan Le Saux.
© 2014 John Wiley & Sons, Inc. Published 2014 by John Wiley & Sons, Inc.

This chapter will occasionally note animal model data to make specific points but primarily focus on human data whenever possible. Further, we will concentrate on mechanisms; briefly review global changes in innate and adaptive immunity with advanced age; outline the intrinsic and extrinsic factors that lead to these changes through immune cell differentiation, exhaustion, and senescence; and close with a brief summary of the clinical implications of the immune senescent phenotype. The data discussed in this chapter summarize immune function in the whole organism; lung-specific changes in host defenses with age are detailed in Chapter 14.

7.2 Barrier defenses and innate immunity in older adults

7.2.1 Barrier defenses

Host defense starts with mucosal integrity, intact skin, and nonspecific host defenses (e.g., stomach acid). Aging is associated with thinning and drying of skin and mucous membranes, achlorhydria, and a number of other changes in nonspecific host defense. This often leads to infection after a smaller number of pathogens than necessary in young adults (e.g., for *Salmonella* spp.) and/or more frequent opportunities for commensal microbes and colonizing pathogens to breach the surface of the skin or mucous membranes. Mucous production and expression of defensins or other secreted antimicrobial substances may also be affected by age, but there are few data [1, 2]. Ciliary function is profoundly affected by comorbidity such as smoking or COPD (e.g., [3]) but also by age. Ciliary beat frequency decreases with age, while nasomucociliary clearance time and ultrastructural abnormalities of cilia increase [4]. In total, these changes in nonspecific defenses render the older adult more likely to experience a breach between the **outside** world and the host whether due to microorganisms that normally inhabit specific niches or acquired pathogens from a dangerous environment (e.g., healthcare-acquired *C. difficile* or methicillin-resistant *Staphylococcus aureus* (MRSA)).

Aging is also associated with changes in the microbiome, at least within the gut. There is a reduction in resident gut anaerobes with age, and these species are replaced by **opportunistic** aerobes such as *Staphylococci* and facultative anaerobes (e.g., Enterobacteriaceae) [5, 6]. The cause(s) for the change in microbiota is not known, but may be related to functional slowing of transit time in the gut, dietary effects, reduced gastric acid, or any one of the myriad physical changes in gut function seen with aging. The change in flora is associated with, though not clearly caused by, increased inflammation in seniors [5].

7.2.2 Innate immunity

Once pathogens gain access by breaching initial nonspecific defenses, the first line of a specific host response is innate immunity – defined as immune functions triggered by recognition of conserved patterns of microbial proteins, lipids, nucleic acids, and other compounds even if the host has never been previously exposed to that specific pathogen. The components of innate immunity include complement and a number of critical cell types: NK cells, polymorphonuclear neutrophils (PMNs), and monocyte-derived cells (macrophages, dendritic cells (DCs)).

Complement levels do not fall and may even slightly increase with age, and minimal data suggest complement-mediated killing is not substantially influenced by age [7]. However, cell responses (e.g., chemotaxis, phagocytosis) to complement triggered actions are often impaired as described later.

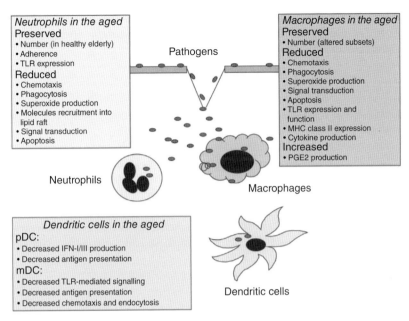

Figure 7.1 Alterations of innate immunity in the aged. Reprinted from Ref. [10] with permission from Elsevier.

There are numerous changes in cell number, type, and response that accompany advancing age as summarized in Figure 7.1. As a complete review is beyond the scope of this chapter, the following paragraphs contain summary information; the reader is referred to [8, 9, 10] for recent reviews.

7.2.2.1 NK cells

NK cell numbers generally increase with age, but the function of these cells is impaired. Antibody-dependent cell cytotoxicity (ADCC) is reduced on a per-cell basis with age despite no change in CD16 expression. However, aging is accompanied by decreasing expression of the activating natural cytotoxicity receptors NKp30 and NKp46 (reviewed in [10]). Although the production of interferon-α by aged NK cells is robust, there is a delayed response with regard to IL-2- and IL-12-driven cytokine and chemokine production [11, 12, 13, 14].

7.2.2.2 Neutrophils

The number of PMNs, their adherence to the epithelium, and the expression of toll-like receptors (TLRs) are unchanged with age. However, chemotaxis, phagocytosis, production of oxidative burst for bacterial killing, recruitment of molecules to lipid rafts, signal transduction, and cell survival/apoptosis are all impaired with advanced age (reviewed in [8, 9, 10]). The effect of age on the production of recently identified neutrophil extracellular traps (NETs) is unknown. The specific impact of any given change in PMN function on infection risk is unknown, but the net effect of cumulative changes within aged neutrophil functions would be predicted to slow initial PMN response to a bacterial challenge and delay recruitment of secondary response cells when compared to young adults. This is supported by observations that older adults presenting with infection frequently have more advanced/severe infection on initial presentation when compared to young adults.

7.2.2.3 Monocyte/macrophages

There is inconsistent data with respect to cytokine production by peripheral blood monocytes or monocyte-derived macrophages depending on whether constitutive or stimulated levels are measured, the type of stimulus used, and whether isolated or mixed cell populations are examined. As noted previously, this chapter focuses on human data which are sparse in this area. However, in a series of very well-done studies, van Duin and colleagues demonstrated that human peripheral blood monocytes from older adults demonstrate lower expression of TLR1 and TLR4 and do not respond to TLR-agonists as well as those of young adults [15]. Further, impaired TLR-triggered responses predicted poor influenza vaccine responses [16], indicating the critical function of these cells to trigger adaptive immune responses may be impaired in older adults.

7.2.2.4 Dendritic cells

There have been a number of studies on the two major subtypes of human DCs, myeloid DC (mDC) and plasmacytoid DC (pDC), but there is no consensus on whether circulating pDC or mDC numbers change with age perhaps due to the very low number of DC in peripheral blood, comparisons between groups of various ages, and use of different markers to identify DC subtypes. However, a recent, extensive examination across the lifespan demonstrated reduced circulating pDC and mDC numbers with age, with the pDC compartment most affected and the greatest change seen between 0–19-year-olds and any of the adult age groups (20–39, 40–59, 60–74), but minimal changes thereafter [17]. At the tissue level, there are data suggesting reduced number and altered morphology of Langerhans cells in the skin and gingiva [8, 18, 19].

Marked decreases in pDC function are seen with advanced age; interferon-α production is reduced after TLR ligand activation or viral infection [20–23]. In contrast, antigen presentation by mDC appears to be intact with age, but migration and pinocytosis/endocytosis are impaired with age [24]. Despite increased IL-6 and TNF-α production after LPS or ssRNA stimulation of myeloid-derived DCs noted by some authors, impaired TLR function across numerous TLRs of native mDC circulating in peripheral blood has been demonstrated and predicts poor response to influenza vaccine similar to the findings regarding peripheral monocytes noted previously [20].

7.3 Adaptive immune responses

7.3.1 B cell number and function

The aged phenotype suggests strongly that B cell dysfunction increases with age. When compared to young adults, older adults generally mount reduced antibody responses after immunization, are more susceptible to pathogens where B cell-mediated defenses play a primary role (e.g., encapsulated organisms such as *Streptococcus pneumoniae*), and experience much higher rates of B cell-derived malignancies (e.g., chronic lymphocytic leukemia, non-Hodgkin's lymphoma; see http://seer.cancer.gov/faststats/). Further, while antibody production is a well-known function of B cells, B cells also play critical effector and regulatory roles in the immune response; these functions are not spared from age-related changes as summarized in the following paragraph.

With advancing age, there is a shift in the bone marrow toward myeloid precursors and away from lymphoid precursors. This may be due to changes in hematopoietic stem cells (HSC) secondary to intrinsic limits with regard to the number of cell divisions (telomere shortening) but also epigenetic changes that are more frequent as one ages and alterations in bone marrow homing (reviewed in [25]). These mechanisms likely underlie a consistently observed, age-related decline in both the percentage of lymphocytes that are CD19+ B cells and their absolute number in peripheral blood (reviewed in [26]).

In addition to declines in typical B cell precursors, recent data suggests there is an age-related accumulation of a novel B cell subset termed "aging-associated B cells" (ABC) [27]. While there are only a few reports regarding ABC function, primarily derived from murine studies (though there seems to be a human correlate of ABC) [27], this B cell subtype appears to arise from mature, naïve B cell precursors after replication driven by TLR and perhaps other innate immune stimuli. ABC are more prevalent in females than males and accumulate at an earlier age in autoimmune-prone mouse strains that mirror accelerated immune aging seen in a number of proinflammatory states. Further, ABC secrete autoantibodies and cytokines and facilitate differentiation of Th17 cells. An accumulation of a specific, senescent B cell subtype with an altered cytokine profile is not surprising and parallels changes noted over many years in the T cell compartment (see section 7.3.2).

While memory responses to antigens after exposure in childhood or young adult life remain robust into late life, responses to new antigens are markedly reduced in late adulthood. The repertoire/diversity of immunoglobulin chains is reduced with age ultimately leading to slower production of antibodies after immunization. It is hypothesized that this is due to reduced T cell help/class switching for many antigens (see section 7.3.2), but even for T cell-independent antigens (e.g., pneumococcal polysaccharide), the antibodies produced in seniors show a reduced capacity to neutralize pathogens or facilitate opsonization despite reaching absolute titers in the circulation similar to those of young adults (reviewed in [25, 26]). This may be due to age-related changes in somatic hypermutation of the immunoglobulin heavy chain, but there are differences between mice and humans that leave this question open at the present time (reviewed in [26, 28]).

7.3.2 T cell number, subtypes, and function

T cells derive their name from the site of primary maturation – the thymus. Thymic atrophy is a well-documented phenomenon of aging, and though its onset predates sexual maturity, the rise in estrogen in females and testosterone in males at puberty parallels accelerated thymic atrophy, and castration can partially reverse and/or delay thymic involution (reviewed in [29]). However, thymic atrophy occurs throughout adulthood despite a general decline in sex steroid levels as one ages. By the time one reaches the sixth decade, most of the thymus is fat with a few islands of lymphoid tissue. The accumulation of adipose within the thymus that may drive thymic involution is suggested by the earlier onset of thymic involution in obese individuals, partial reversal by leptin administration, and acceleration of thymic atrophy after transplantation of adipose tissue to the thymic bed (reviewed in [29]). Adipose tissue is proinflammatory and profibrotic via secretion of a number of cytokines, and thus, fat tissue itself may be a major player in reducing thymic emigrants with age.

Thymic atrophy is likely a major cause of the marked decline in the absolute number of naïve T cells as one ages, though alternate niches of T cell maturation can keep up for a time [30]. The loss of thymic emigrants means the naïve T cell pool must be replenished by autoproliferation of naïve T cells in the periphery. It is estimated that 0.1%–1% (or about 3×10^8–10^9 T cells) must be replaced daily in older adults [31]. Further, the T cell repertoire contracts with age – there is relatively preserved T cell receptor (TCR) diversity through the seventh decade of life, but after age 70, there is a substantial contraction (Figure 7.2) [31]. This relative decline in TCR diversity is thought to result in a much narrower range of antigens one can respond to when challenged. These same age-related changes in T cell pools/diversity and dependence of autoproliferation of peripheral T cell pools have been documented in nonhuman primates as a model of human aging [32]. Further, inflammatory stimuli appear to accelerate turnover of peripheral T cell pools in a nonspecific fashion, perhaps limiting the predetermined lifespan of naïve T cells [32].

At the same time naïve T cells are declining, memory T cells are accumulating due to a lifetime of antigen exposure. The percentage of T cells of memory phenotype expands rapidly in the first few decades of life and then more slowly over time as one ages into the latter decades [31]. For chronic or repeated infections, this expansion becomes progressively oligoclonal and can lead to immune **exhaustion**, particularly in response to specific viruses (see section 7.3.3).

7.3.3 T cell activation, differentiation, exhaustion, and senescence

After T cells are exposed to antigen, they become activated and differentiate into effector and memory phenotypes. Memory T cells must last a lifetime, and similar to the need for naïve cells to autoproliferate to replenish the naïve T cell pool after thymic involution, autoproliferation of memory T cells is necessary for homeostasis of the memory T cell pool [32, 33].

The maintenance and homeostatic turnover of competent T cell pools are affected by many intrinsic and extrinsic factors. One critical extrinsic factor for memory T cells is repeated antigen stimulation (reviewed in [34, 35]). This is probably best illustrated by viral infections that can be *acute* with resolution (e.g., measles, influenza), or *persistent*, which must be divided into *chronic* with constant viral replication (e.g., HIV, hepatitis C), or *cyclic* with latency/reactivation (e.g., HSV, CMV) [36]. In cyclic infections, restimulation of memory T cells by antigen in a milieu of chronic, low-level inflammation leads to clonal expansion of T cells, particularly CD8+ T cells [34, 37, 38]. The accumulated CD8+ T cell clones are typically described in the literature as **exhausted** and **senescent** [34, 36, 37, 38, 39]. These two terms are often used interchangeably with regard to aging T cells but are distinct [35]. Replicative senescence refers to cells that no longer divide in response to typical stimuli (reviewed in [35]). Characteristically, human senescent cells have short telomeres with minimal telomerase activity. A review of telomere length and telomerase activity is well beyond the scope of this chapter, but it is known that telomeres of immune cells are shorter and telomerase activity is reduced in older adults when compared to young adults (reviewed in [40]). Age-related telomere dysfunction will be covered in Chapter 4.

Telomere length and telomerase activity have been often considered the **gold standard** markers of cellular senescence. More recently, **senescent** cells have been identified by the expression of cell cycle arrest proteins that inhibit cell division. One such protein is p16ink4a (p16) (reviewed in [41]), a robust biomarker of senescence in most cells, including T cells [42, 43, 44, 45]. Like telomere length, p16 expression is not age specific as it can be affected by environmental stimuli [46] including oxidative stress. Whereas telomere length only declines by a factor of about two, T cell p16 expression increases about 20-fold as one ages, though it is not linear with accelerated increases in early/mid adulthood and a slower rate in mid- to late adult years [42].

Shortened telomeres and p16 are both associated with cell cycle arrest and therefore help define **replicative** senescence. In contrast to cell senescence, T cell exhaustion refers to a change in T cell phenotype during successive rounds of restimulation [36]. Exhaustion is seen mostly with truly continuous antigenic stimulation (e.g., as seen with HIV and hepatitis C that do not enter a latent phase, but are continuously producing virus). T cell exhaustion is identified by alterations in cell surface markers (↑ CD57, PD-1, Tim-3, BIM, BLIMP-1 and ↓ CD28, CTLA-4) but characterized by specific changes in effector function (↓ IL-2, ↑ cyto-toxicity, and expression of inflammatory cytokines) [35].

In contrast to chronic viral infections, persistent latent infections (e.g., herpesviruses) result in periodic antigen exposure that produces a different phenotype [34]. Large pools of CD8 cells accumulate with age in response to systemic infection with herpesvirus, and a fraction of those cells do not proliferate; however, most, if not all, do retain function deep into advanced age [47, 48].

T cell replicative senescence and exhaustion are likely part of the same continuum, and most **exhausted** CD8+ T cells identified by surface markers are also **senescent** in that they have shortened telomeres and do not proliferate after stimulation [35, 39]. While many investigators believe senescence and exhaustion to be **end stage**, some aspects may be reversible. Telomerase expression can restore proliferation [49, 50]. Further, expression of exhaustion-related T cell inhibitory proteins on the cell surface triggers senescence [35], but blocking those inhibitory proteins (e.g., PD-1) can restore antigen-specific T cell replication and functional responses [51], though not telomerase activity [52].

Immune responses wane in seniors through exhaustion and senescence, but compensatory changes in other immune compartments attempt to **backfill** host defenses. One example is the upregulation of KIR/CD158 on αβ T cells [37]. Other adaptive responses may be attempts to avoid accumulation of even greater numbers of a few clones – for example, those directed versus CMV [34].

7.4 Consequences of immune senescence

7.4.1 Impaired vaccine responses, increased risk of infection, and age-related illness

Impaired vaccine responses are a major consequence of immune senescence that bears relevance for the aging lung. In general, recall responses to booster vaccines are robust well into advanced age, at least into one's 60s, but appear to wane above age 70 consistent with the drop in T cell diversity (Figure 7.2 (reviewed in [53, 54]). For example, the Shingles Prevention Study demonstrated strong efficacy for the prevention of clinical zoster in those aged 60–69, but reduced efficacy for the prevention of overt disease for those aged 70 and over, though the vaccine did offer reduction in the severity if not incidence of disease in the older age group [55].

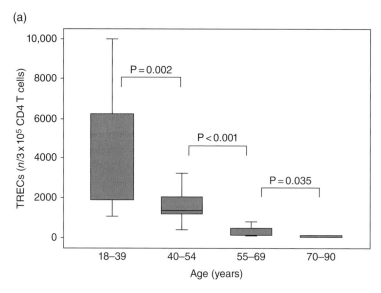

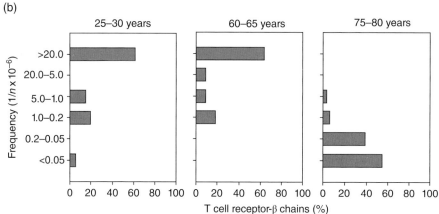

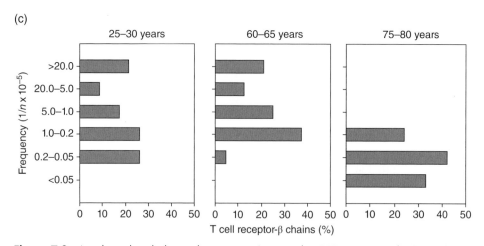

Figure 7.2 Age-dependent decline in thymic output (measured as TCR excision circles (TRECs)) (a) and TCR β-chain diversity in naïve (b) and memory (c) CD4+ T cells in different age groups. With permission from Ref. [31]. Copyright 2005, The American Association of Immunologists, Inc.

Older adults are also at much greater risk than young adults for infection-related illness. There are many publications documenting the increased risk of illness and/or severe illness with advanced age in sepsis, pneumonia, influenza, urinary tract infection, skin and soft-tissue infection, and many other infectious diseases. To more clearly illustrate the implications of specific immune changes with regard to memory and naïve responses highlighted previously, zoster again provides an excellent example. Zoster risk increases markedly from about 2/1000 person years for 25–34-year-olds to 15/1000 person years in those aged 75+ years [56]. Clinical zoster of course represents waning immunity to a virus most of us have been exposed to as children that is latent in the dorsal root and cranial nerve ganglia. Although zoster can occur in young adults, severe sequelae rarely occur until one reaches 70+ years, remarkably correlating with the decline in T cell diversity about the same age (Figure 7.2). However, just as with vaccine responses to neoantigens are impaired in old age, exposure to new pathogens at an advanced age is much more likely to result in poor immunity and thereby adverse outcomes. An excellent example is West Nile virus in which both the incidence of disease and mortality increase 30-fold for those ≥90 years versus those <40.

Recent evidence also suggests that immune senescence leads to impaired resolution of inflammation after an infectious event. For example, the initial degree of systemic inflammation is similar in young or old adults presenting with documented *S. pneumoniae* bacteremia; however, inflammatory markers decline much more slowly in seniors [57], suggesting that immune dysfunction either leads to more robust/prolonged proinflammatory signals or to impaired anti-inflammatory pathways after a stimulus. This is a critical finding because low-grade inflammation may contribute to slow/steady progression of chronic illnesses (e.g., atherosclerosis), but there is also evidence that acute inflammation induced by infection, but not more trivial stimuli (i.e., vaccines), can lead to a critical window of risk for adverse events. Inflammation-triggered hypercoagulation likely underlies the increased risk of myocardial infarction and stroke postinfection [58] and has been linked to p16 expressing (i.e., senescent) cells [59].

7.4.2 Immune senescence: A cause of aging itself

Senescent cells develop a specific phenotype including a profoundly altered gene expression profile called the "senescence-associated secretory phenotype" (SASP) [41]. SASP reprogramming typically results in secretion of proinflammatory cytokines. One example is highly differentiated effector memory T cells, identified by the surface markers CD8+CD57+CD28 in humans, accumulate with aging and produce proinflammatory cytokines (IL-6, TNF-α) upon stimulation *ex vivo* (reviewed in [60, 61, 62]). However, whether they do so *in vivo*, and whether they must encounter antigen to do so, constitutively produce cytokines, or whether they do so upon subthreshold stimulation perhaps after exposure to cross-reactive antigen remains to be established. Moreover, while it is generally agreed that the proinflammatory state associated with aging strongly correlates with, and is hypothesized to be a *cause* of, age-related illness (reviewed in [63]), the source of this systemic inflammation in seniors remains unresolved at this time. Indeed, while constitutive secretion of cytokines from differentiated effector memory cells (with or without **senescent** characteristics) is one plausible candidate, other viable candidates include accumulating proinflammatory adipose tissue and changes in mucosal barrier permeability and/or microbial colonization.

There is strong evidence that immune senescence is an important contributor to immune dysfunction, illness, and disability in seniors, however. As noted previously, reductions in naïve T cells with age, poor T cell help, and reduced TCR repertoire diversity all likely contribute to poor immune responses in older adults [64], but there is also evidence of **bystander effects** of senescent cells hypothesized to be due to cytokine secretion. For example, influenza vaccine responses do not correlate with the loss of naïve CD4 cells or the percentage of memory cells versus naïve cells, but do strongly correlate with the percentage of senescent CD8 cells (identified by the loss of CD28 expression [65, 66, 67, 68]). How accumulation of senescent T cells might lead to poor vaccine responses is not clear and has been hypothesized to be due to SASP of these cells. However, to influence priming, senescent T cells would have to affect lymph nodes, end organs, or other sites that affect immunity, and so far there is no experimental evidence for this.

The speculation that the inflammatory milieu promoted by senescent cells contributes to aging itself – the so-called inflamm-aging (reviewed in [63, 69]) – may be amplified in those with persistent viral infections [61, 70]. According to this theory, persistent infection and inflammation are interrelated drivers of aging immunity, and senescent T cells and inflammation from uncontrolled viral infection combine to accelerate loss of homeostatic mechanisms and premature **age** of immune cells (Figure 7.2) [38, 62], but this remains a hypothesis to be tested.

One very interesting animal model lends support to the notion that SASP contributes to aging. Baker et al. [71] demonstrated that the elimination of senescent cells in a progeroid mouse model by deleting p16 expressing (i.e., senescent) cells resulted in marked improvement of the accelerated age phenotype in those mice. The effect was presumably achieved through elimination of the senescent cells and improved function of surrounding cells previously impaired by SASP [72]. The expression system used by Baker et al. [71] utilized an approach focused primarily on eliminating p16 expressing adipocytes, but all cells in which the p16 promoter was active, including senescent T cells, were likely to be affected. However, an increase in maximal lifespan – the gold standard for antiaging interventions – was not demonstrated in that study. Further evidence that senescent T cells might play a direct causal role in aging comes from experiments in which p16 was inactivated in either T or B cell lineages **at a specific age** using a *Lck-Cre* system [73]. The inactivation of p16 resulted in maintenance of a **youthful** phenotype when inactivated in the T cell compartment but led to enhanced lymphoid malignancies in B cells.

7.5 Conclusion

The immune system undergoes dramatic changes with advancing age in all compartments of immunity. This generally results in an immune senescent phenotype characterized by low-level constitutive inflammation, impaired mucosal defenses, reduced numbers of naïve T and B cells and expanded memory cells, reduced diversity of T and B cell responses, and poor effector function when compared to young adults. The sequelae of these changes result in poor vaccine responses, increased susceptibility to infection, and increased risk of infection-triggered exacerbation of underlying illnesses (e.g., stroke, myocardial infarction, chronic lung diseases). The degree to which systemic immune senescence contributes to illness and infection in the lung is specifically addressed in Chapter 14 (KC Meyer), but evidence in animal models and humans suggests that no organ system is spared. Immune senescence may also contribute to aging itself, a hypothesis being tested in various translational models of aging.

References

1. Harimurti K, Djauzi S, Witarto AB, Dewiasty E. Human beta-defensin 2 concentration of respiratory tract mucosa in elderly patients with pneumonia and its associated factors. *Acta Med Indones* 2011;43:218–223.
2. Matsuzaka K, Sato D, Ishihara K, et al. Age-related differences in localization of beta-defensin-2 in human gingival epithelia. *Bull Tokyo Dent Coll* 2006;47:167–170.
3. Sommer JU, Stuck BA, Heiser C, Kassner SS, Hormann K, Sadick H. Ciliary function of the nose in patients with Osler's disease and the effect of topically applied estrogens as a nose ointment. *Rhinology* 2011;49:407–412.
4. Ho JC, Chan KN, Hu WH, et al. The effect of aging on nasal mucociliary clearance, beat frequency, and ultrastructure of respiratory cilia. *Am J Respir Crit Care Med* 2001;163:983–988.
5. Biagi E, Candela M, Franceschi C, Brigidi P. The aging gut microbiota: New perspectives. *Ageing Res Rev* 2011;10:428–429.
6. Tiihonen K, Ouwehand AC, Rautonen N. Human intestinal microbiota and healthy ageing. *Ageing Res Rev* 2010;9:107–116.
7. Simell B, Vuorela A, Ekstrom N, et al. Aging reduces the functionality of anti-pneumococcal antibodies and the killing of *Streptococcus pneumoniae* by neutrophil phagocytosis. *Vaccine* 2011;29:1929–1934.
8. Shaw AC, Joshi S, Greenwood H, Panda A, Lord JM. Aging of the innate immune system. *Curr Opin Immunol* 2010;22:507–513.
9. Panda A, Arjona A, Sapey E, et al. Human innate immunosenescence: Causes and consequences for immunity in old age. *Trends Immunol* 2009;30:325–333.
10. Solana R, Tarazona R, Gayoso I, Lesur O, Dupuis G, Fulop T. Innate immunosenescence: Effect of aging on cells and receptors of the innate immune system in humans. *Semin Immunol* 2012;24:331–341.
11. Le Garff-Tavernier M, Beziat V, Decocq J, et al. Human NK cells display major phenotypic and functional changes over the life span. *Aging Cell* 2010;9:527–535.
12. Mariani E, Meneghetti A, Neri S, et al. Chemokine production by natural killer cells from nonagenarians. *Eur J Immunol* 2002;32:1524–1529.
13. Mariani E, Pulsatelli L, Meneghetti A, et al. Different IL-8 production by T and NK lymphocytes in elderly subjects. *Mech Ageing Dev* 2001;122:1383–1395.
14. Mariani E, Pulsatelli L, Neri S, et al. RANTES and MIP-1alpha production by T lymphocytes, monocytes and NK cells from nonagenarian subjects. *Exp Gerontol* 2002;37:219–226.
15. van DD, Mohanty S, Thomas V, et al. Age-associated defect in human TLR-1/2 function. *J Immunol* 2007;178:970–975.
16. van DD, Allore HG, Mohanty S, et al. Prevaccine determination of the expression of costimulatory B7 molecules in activated monocytes predicts influenza vaccine responses in young and older adults. *J Infect Dis* 2007;195:1590–1597.
17. Orsini G, Legitimo A, Failli A, Massei F, Biver P, Consolini R. Enumeration of human peripheral blood dendritic cells throughout the life. *Int Immunol* 2012;24:347–356.
18. Bodineau A, Coulomb B, Folliguet M, et al. Do Langerhans cells behave similarly in elderly and younger patients with chronic periodontitis? *Arch Oral Biol* 2007;52:189–194.
19. Zavala WD, Cavicchia JC. Deterioration of the Langerhans cell network of the human gingival epithelium with aging. *Arch Oral Biol* 2006;51:1150–1155.
20. Panda A, Qian F, Mohanty S, et al. Age-associated decrease in TLR function in primary human dendritic cells predicts influenza vaccine response. *J Immunol* 2010;184:2518–2527.
21. Shodell M, Siegal FP. Circulating, interferon-producing plasmacytoid dendritic cells decline during human ageing. *Scand J Immunol* 2002;56:518–521.
22. Jing Y, Shaheen E, Drake RR, Chen N, Gravenstein S, Deng Y. Aging is associated with a numerical and functional decline in plasmacytoid dendritic cells, whereas myeloid dendritic cells are relatively unaltered in human peripheral blood. *Hum Immunol* 2009;70:777–784.
23. Sridharan A, Esposo M, Kaushal K, et al. Age-associated impaired plasmacytoid dendritic cell functions lead to decreased CD4 and CD8 T cell immunity. *Age (Dordr)* 2011;33:363–376.
24. Agrawal A, Agrawal S, Cao JN, Su H, Osann K, Gupta S. Altered innate immune functioning of dendritic cells in elderly humans: A role of phosphoinositide 3-kinase-signaling pathway. *J Immunol* 2007;178:6912–6922.

25. Kogut I, Scholz JL, Cancro MP, Cambier JC. B cell maintenance and function in aging. *Semin Immunol* 2012;24:342–349.

26. Bulati M, Buffa S, Candore G, et al. B cells and immunosenescence: A focus on IgG+IgD−. *Ageing Res Rev* 2011;10:274–284.

27. Rubtsov AV, Rubtsova K, Fischer A, et al. Toll-like receptor 7 (TLR7)-driven accumulation of a novel CD11c(+) B-cell population is important for the development of autoimmunity. *Blood* 2011;118:1305–1315.

28. Frasca D, Blomberg BB. Aging affects human B cell responses. *J Clin Immunol* 2011;31:430–435.

29. Dooley J, Liston A. Molecular control over thymic involution: From cytokines and microRNA to aging and adipose tissue. *Eur J Immunol* 2012;42:1073–1079.

30. Ratts RB, Weng NP. Homeostasis of lymphocytes and monocytes in frequent blood donors. *Front Immunol* 2012;3:271.

31. Naylor K, Li G, Vallejo AN, et al. The influence of age on T cell generation and TCR diversity. *J Immunol* 2005;174:7446–7452.

32. Cicin-Sain L, Messaoudi I, Park B, et al. Dramatic increase in naive T cell turnover is linked to loss of naive T cells from old primates. *Proc Natl Acad Sci U S A* 2007;104:19960–19965.

33. Nikolich-Zugich J, Li G, Uhrlaub JL, Renkema KR, Smithey MJ. Age-related changes in CD8 T cell homeostasis and immunity to infection. *Semin Immunol* 2012;24:356–364.

34. Nikolich-Zugich J. Ageing and life-long maintenance of T-cell subsets in the face of latent persistent infections. *Nat Rev Immunol* 2008;8:512–522.

35. Akbar AN, Henson SM. Are senescence and exhaustion intertwined or unrelated processes that compromise immunity? *Nat Rev Immunol* 2011;11:289–295.

36. Virgin HW, Wherry EJ, Ahmed R. Redefining chronic viral infection. *Cell* 2009;138:30–50.

37. Vallejo AN, Mueller RG, Hamel DL Jr, et al. Expansions of NK-like alphabetaT cells with chronologic aging: Novel lymphocyte effectors that compensate for functional deficits of conventional NK cells and T cells. *Ageing Res Rev* 2011;10:354–361.

38. Weng NP, Akbar AN, Goronzy J. CD28(−) T cells: Their role in the age-associated decline of immune function. *Trends Immunol* 2009;30:306–312.

39. Brunner S, Herndler-Brandstetter D, Weinberger B, Grubeck-Loebenstein B. Persistent viral infections and immune aging. *Ageing Res Rev* 2011;10:362–369.

40. Weng NP. Telomeres and immune competency. *Curr Opin Immunol* 2012;24:470–475.

41. Coppe JP, Desprez PY, Krtolica A, Campisi J. The senescence-associated secretory phenotype: The dark side of tumor suppression. *Annu Rev Pathol* 2010;5:99–118.

42. Liu Y, Sanoff HK, Cho H, et al. Expression of p16(INK4a) in peripheral blood T-cells is a biomarker of human aging. *Aging Cell* 2009;8:439–448.

43. Krishnamurthy J, Torrice C, Ramsey MR, et al. Ink4a/Arf expression is a biomarker of aging. *J Clin Invest* 2004;114:1299–1307.

44. Coppe JP, Rodier F, Patil CK, Freund A, Desprez PY, Campisi J. Tumor suppressor and aging biomarker p16(INK4a) induces cellular senescence without the associated inflammatory secretory phenotype. *J Biol Chem* 2011;286:36396–36403.

45. Connoy AC, Trader M, High KP. Age-related changes in cell surface and senescence markers in the spleen of DBA/2 mice: A flow cytometric analysis. *Exp Gerontol* 2006;41:225–229.

46. Song Z, von FG, Liu Y, et al. Lifestyle impacts on the aging-associated expression of biomarkers of DNA damage and telomere dysfunction in human blood. *Aging Cell* 2010;9:607–615.

47. Lang A, Nikolich-Zugich J. Functional CD8 T cell memory responding to persistent latent infection is maintained for life. *J Immunol* 2011;187:3759–3768.

48. Cicin-Sain L, Sylwester AW, Hagen SI, et al. Cytomegalovirus-specific T cell immunity is maintained in immunosenescent rhesus macaques. *J Immunol* 2011;187:1722–1732.

49. Akbar AN, Soares MV, Plunkett FJ, Salmon M. Differential regulation of CD8+ T cell senescence in mice and men. *Mech Ageing Dev* 2000;121:69–76.

50. Fauce SR, Jamieson BD, Chin AC, et al. Telomerase-based pharmacologic enhancement of antiviral function of human CD8+ T lymphocytes. *J Immunol* 2008;181:7400–7406.

51. Di MD, Azevedo RI, Henson SM, et al. Reversible senescence in human CD4+CD45RA+. *J Immunol* 2011;187:2093–2100.

52. Henson SM, Macaulay R, Franzese O, Akbar AN. Reversal of functional defects in highly differentiated young and old CD8 T cells by PDL blockade. *Immunology* 2012;135:355–363.

53. High KP, D'Aquila RT, Fuldner RA, et al. Workshop on immunizations in older adults: Identifying future research agendas. *J Am Geriatr Soc* 2010;58:765–776.

54. High K. Immunizations in older adults. *Clin Geriatr Med* 2007;23:669–685.
55. Oxman MN, Levin MJ, Johnson GR, et al. A vaccine to prevent herpes zoster and postherpetic neuralgia in older adults. *N Engl J Med* 2005;352:2271–2284.
56. Donahue JG, Choo PW, Manson JE, Platt R. The incidence of herpes zoster. *Arch Intern Med* 1995;155:1605–1609.
57. Bruunsgaard H, Skinhoj P, Qvist J, Pedersen BK. Elderly humans show prolonged in vivo inflammatory activity during pneumococcal infections. *J Infect Dis* 1999;180:551–554.
58. Smeeth L, Thomas SL, Hall AJ, Hubbard R, Farrington P, Vallance P. Risk of myocardial infarction and stroke after acute infection or vaccination. *N Engl J Med* 2004;351:2611–2618.
59. Cardenas JC, Owens AP III, Krishnamurthy J, Sharpless NE, Whinna HC, Church FC. Overexpression of the cell cycle inhibitor p16INK4a promotes a prothrombotic phenotype following vascular injury in mice. *Arterioscler Thromb Vasc Biol* 2011;31:827–833.
60. Onyema OO, Njemini R, Bautmans I, Renmans W, De WM, Mets T. Cellular aging and senescence characteristics of human T-lymphocytes. *Biogerontology* 2012;13:169–181.
61. Dock JN, Effros RB. Role of CD8 T cell replicative senescence in human aging and in HIV-mediated immunosenescence. *Aging Dis* 2011;2:382–397.
62. Macaulay R, Akbar AN, Henson SM. The role of the T cell in age-related inflammation. *Age (Dordr)* 2013;35(3):563–572.
63. Larbi A, Franceschi C, Mazzatti D, Solana R, Wikby A, Pawelec G. Aging of the immune system as a prognostic factor for human longevity. *Physiology (Bethesda)* 2008;23:64–74.
64. Maue AC, Yager EJ, Swain SL, Woodland DL, Blackman MA, Haynes L. T-cell immunosenescence: Lessons learned from mouse models of aging. *Trends Immunol* 2009;30:301–305.
65. Schneider A, Zimmer HG. Effect of inosine on function and adenine nucleotide content of the isolated working rat heart: Studies of postischemic reperfusion. *J Cardiovasc Pharmacol* 1991;17:466–473.
66. Saurwein-Teissl M, Lung TL, Marx F, et al. Lack of antibody production following immunization in old age: Association with CD8(+)CD28(−) T cell clonal expansions and an imbalance in the production of Th1 and Th2 cytokines. *J Immunol* 2002;168:5893–5899.
67. Xie D, McElhaney JE. Lower GrB+ CD62Lhigh CD8 TCM effector lymphocyte response to influenza virus in older adults is associated with increased CD28null CD8 T lymphocytes. *Mech Ageing Dev* 2007;128:392–400.
68. Trzonkowski P, Mysliwska J, Szmit E, et al. Association between cytomegalovirus infection, enhanced proinflammatory response and low level of anti-hemagglutinins during the anti-influenza vaccination – an impact of immunosenescence. *Vaccine* 2003;21:3826–3836.
69. Franceschi C, Capri M, Monti D, et al. Inflammaging and anti-inflammaging: A systemic perspective on aging and longevity emerged from studies in humans. *Mech Ageing Dev* 2007;128:92–105.
70. Wills M, Akbar A, Beswick M, et al. Report from the second cytomegalovirus and immunosenescence workshop. *Immun Ageing* 2011;8:10.
71. Baker DJ, Wijshake T, Tchkonia T, et al. Clearance of p16Ink4a-positive senescent cells delays ageing-associated disorders. *Nature* 2011;479:232–236.
72. Peeper DS. Ageing: Old cells under attack. *Nature* 2011;479:186–187.
73. Liu Y, Johnson SM, Fedoriw Y, et al. Expression of p16(INK4a) prevents cancer and promotes aging in lymphocytes. *Blood* 2011;117:3257–3267.
74. Girard TD, Opal SM, Ely EW. Insights into severe sepsis in older patients: From epidemiology to evidence-based management. *Clin Infect Dis* 2005;40:719–727.

8 Developmental and Physiological Aging of the Lung

Kent E. Pinkerton[1], Lei Wang[1], Suzette M. Smiley-Jewell[1], Jingyi Xu[1,2], and Francis H.Y. Green[3]

[1] Center for Health and the Environment, University of California Davis, Davis, California, USA
[2] Affiliated Zhongshan Hospital of Dalian University, Dalian, China
[3] University of Calgary, Calgary, Alberta, Canada

8.1 Introduction

Lung development and aging is a continuum that begins *in utero* and continues throughout life. While many studies have examined the anatomy and physiology of the developing lung, fewer studies have evaluated these parameters in the aging lung. This information is needed due to the rapid growth in the number of elderly individuals in developed countries; a growing number of studies demonstrate that the aging pulmonary system (>65 years) are at increased risk for adverse health effects from such factors as environmental insults, pathogens, and aeroallergens [1, 2, 3]. It is generally agreed that lung function declines with age and that oxidant stress related to smoking, chronic lung inflammation, or cardiopulmonary diseases may increase the rate of decline [1, 4]. These physiological changes are accompanied by structural changes in the lung. A brief overview of the aging lung is provided in this chapter along with information on how the lung changes with age in the rat, which due to its size and lifespan is ideal for studies of the maturing lung.

8.2 The aging lung

8.2.1 Alterations in lung function and anatomy

The human lung takes until 20–25 years of age to fully mature. Thereafter, aging is associated with a progressive decline in lung function [5, 6]. Dead space ventilation increases with age [5], and the transport of respiratory gases across the alveolar wall measured by the pulmonary transfer factor for carbon monoxide (TLCO) declines with age [7]; both adversely affect oxygen tension. Age-related changes in cardiovascular function contribute to these effects [8] as do aging effects in the central and peripheral nervous system. Older adults have a decreased sensation of dyspnea and ventilatory response to hypoxia and hypercapnia [9, 10] accompanied by a decrease in CO_2 threshold [11].

Molecular Aspects of Aging: Understanding Lung Aging, First Edition. Edited by Mauricio Rojas, Silke Meiners and Claude Jourdan Le Saux.

Even in the absence of disease, the respiratory system undergoes a variety of anatomical and physiological changes with age (Tables 8.1 and 8.2) [8, 12]. Structural changes include deformities of the chest wall and thoracic spine that impair respiratory system compliance and increase the work of breathing [13]. With age, there is a diminished

Table 8.1 Age-related structural changes in the chest wall, diaphragm, and lung.

Chest wall and diaphragm
- Decreased chest wall compliance
  - Changes in costochondral cartilage
  - Spinal osteoporosis
  - Spinal osteoarthritis
- Decreased respiratory muscle strength
  - Loss in muscle protein mass
- Increased visceral fat

Lung parenchyma
- Altered alveolar matrix proteins
- Senile emphysema

Airways
- Changes in airway receptors
- Calcification of airway cartilage

Pulmonary vasculature
- Remodeling of elastic and smooth muscle of pulmonary arteries

Accumulated genetic injury

Table 8.2 Physiological changes in lung function in the elderly.

Dynamic flow	
• FEV_1	Decreased
• FVC	Decreased
• FEV_1/FVC ratio	Decreased
Lung volumes	
• TLC	Unchanged
• VC	Decreased
• RV	Increased
• FRC	Increased
Gas exchange	
• DLCO/Va	Decreased
• Minute ventilation (MV)	Unchanged
• Dead space ventilation	Increased
Exercise capacity	
• VO_2 max	Decreased
Compromised defense mechanisms	
• Antioxidants	Decreased
• Immune cell function	Variable
• Neural reflexes	Decreased
• Mucociliary clearance	Decreased
Response to hypoxia and hypercapnia	Decreased
• Sleep apnea	Increased

Adapted from Green and Pinkerton (2004) and Sharma and Goodwin (2006).
FEV1, forced vital expiratory volume in one second; FVC, forced vital capacity; TLC, total lung capacity; VC, vital capacity; RV, residual volume; FRC, functional residual capacity.

respiratory effort in response to upper airway occlusion and impaired perception of bronchoconstriction [7]. Moreover, respiratory muscle strength decreases and can impair effective cough, which is important in airway clearance [14]. The lung parenchyma loses supporting structure, leading to dilation of air spaces or **senile emphysema** [15]. An age-related decrease in elastic fibers with an increase in type III collagen deposition within alveolar walls is indicative of an impaired repair mechanism, which could play a significant role in the pathogenesis of senile emphysema [16, 17]. Lastly, adult aging *per se* does not appear to alter the regional deposition fraction of aerosols [18]; however, changes in inspiratory peak flow may impair deposition [5].

8.2.2 Oxidative stress and lung antioxidant defenses

A commonly held theory of aging is based on the free radical or oxidative stress hypothesis: the accumulation of reactive oxygen species (ROS) in cells with increasing age results in various forms of reversible and irreversible oxidative modifications of proteins (carbonylation or nitromodifications), lipids (hydroperoxide lipid derivatives), and DNA (adducts and breaks) that eventually lead to loss of molecular function. The human lung is continuously exposed to oxidative stress. Particulates, such as cigarette smoke, ambient airborne aerosols, coal dust, or silica, all possess intrinsic free radical activity [19]. Diesel emission particulates also contain compounds that catalyze the generation of ROS [20], as well as other exogenous toxins that activate lung resident cells to produce ROS [21].

The lung has a remarkable array of responses and defenses to oxidative stress. Low-level stress may invoke adaptive responses, including growth arrest and preferential expression of genes responsible for repair of damage [22, 23]. The lung normally defends itself against ROS damage through the use of specific ROS-reducing mechanisms that encompass enzymatic reactions involving superoxide dismutase (SOD), glutathione peroxidases, and catalases or nonenzymatic components such as vitamin A, C, and E; urate; ferritin; ceruloplasmin; surfactant; and ubiquinone [24].

As the lung ages, change in antioxidant defenses occurs such as decreased ascorbic acid [25, 26] and glutathione levels [27, 28] and decreased activity of SOD and glutathione peroxidase [29] in lung tissues. Nuclear factor erythroid-derived 2 (NRF2), a transcription factor that protects cells and tissues from oxidative stress by activating protective antioxidant and detoxifying enzymes [30], has been found decreased in the alveolar macrophages of older current smokers and patients with COPD compared with younger subjects [31].

8.2.3 Immune system changes with aging

Immune mechanisms underlie many chronic diseases of the lung. Both innate immunity and acquired immunity are affected by aging [12], as is discussed in detail in Chapter 7. The impact of age on innate and adaptive immune responses includes the functions of antigen-presenting cells, phagocytic cells, inflammatory mediators released from leukocytes, natural killer lymphocytes, antimicrobial molecules (such as nitric oxide), surfactant-associated proteins, defensins, lactoferrin, and complement [12, 32].

Changes in the composition of bronchoalveolar lavage fluid have been reported in the elderly versus the young (Table 8.3) [12, 32]. These parameters reflect innate as well as acquired

Table 8.3 Changes in bronchoalveolar lavage fluid in healthy aged individuals.

Parameter	Change
Lymphocytes	↑
CD4+/CD8 + T-cell ratio	↑
HLA-DR + T-cells	↑
B-cells	↓
IgM, IgA, and IgG	↑
Total protein	↑
Neutrophils	↑
Interleukins (IL-6, IL-8)	↑
α1-antitrypsin	↑

Adapted from Meyer (2001) and Green and Pinkerton (2004).

immune responses. The gradual reduction in the adaptive immune response is associated with an innate immune system stimulated and upregulated by external environmental insults or internal antigenic stimuli from oxidative stress, creating a chronic proinflammatory state with aging. Inflammation is central to the pathogenesis of obstructive lung diseases, including asthma, chronic bronchitis, and emphysema [12].

It is unclear how much aging of the respiratory system increases susceptibility of the elderly to respiratory infections, particularly those caused by newly emerging and reemerging pathogens. No doubt reduced antioxidant capacity, suboptimal nutrition, deteriorating lung mechanics, and declined immunity all play a role in elevated susceptibility [12]. In the case of influenza infections, even though the hospitalization rates for children less than 5 years and adults over 70 years of age are almost identical, individuals older than 70 years have a 35-fold increase in mortality [33]. Vaccination can reduce the rates of hospitalization; however, protection induced by immunizations is diminished in the elderly compared to adults as demonstrated by lower antibody titers and higher rates of respiratory illness [34]. In addition, cell-mediated immune responses to vaccinations are decreased in the elderly [35], and increased morbidity and mortality due to influenza occurs as a result of secondary bacterial and viral infections [36].

8.2.4 Body mass

Body mass is positively associated with airway hyperresponsiveness [37] and with progressive reductions in FVC, FEV_1, and FEV_1/FVC ratio [38]. Cross-sectional [39, 40] and longitudinal [41] studies have shown a relationship between asthma and obesity, particularly in women [42]. Older individuals with obstructive lung disease were found to have excessive abdominal visceral fat and increased plasma IL-6 compared to older individuals with normal lung function [43]. Static lung volumes significantly increase following weight loss in middle-aged and older, obese men. In contrast, aerobic exercise has no effect on pulmonary function but does increase maximal oxygen uptake (VO_2 max) [44]. The contribution of body composition, physical activity, and smoking to lung function in older people has been investigated [45]; fat-free mass and physical activity both exerted significant independent effects on FEV_1. These results, in contrast to the former study [44], indicate that heavy intense physical activity may be more important in contributing to forced expiratory function than previously recognized [12].

8.2.5 Airway receptor and endocrine changes with aging

Age-related changes in airway receptors, critical for airway homeostasis, are not well understood [8]. Changes in muscarinic receptor subtypes and receptor coupling to G proteins have been noted with senescence [46], which might account for an increase in bronchial responsiveness to methacholine challenge in the elderly [47] and to increased responsiveness to environmental exposures in older mice [48]. Response to β agonists is also impaired in older subjects [49]. While β-adrenoreceptor density remains unchanged with age, however, receptor affinity is significantly reduced [50]. Cysteinyl leukotriene (CysLT1) receptors undergo functional changes with age and are less likely to respond to drugs that are effective in younger individuals to treat the same disorders [51, 52]. Studies in animals show that substance P receptor (neurokinin-1 receptor) decreases with age in normal sheep [53].

Hormonal changes are thought to play a role for the loss of lung tissue mass, impaired respiratory and peripheral muscle function, and reduced exercise capacity in older individuals. Changes in the hypothalamic pituitary–adrenal axis occur with age, and attenuated diurnal variability may account for some hormonal changes [54]. Growth hormone (GH), insulin-like growth factor [55], estrogen [56], glucocorticoids [57], testosterone [58], and vitamin D [59] have been demonstrated to change with increasing age. An age-related decline in GH parallels changes in body composition, such as a reduction in lean body mass and bone mineral density and increased visceral fat [55]. Loss in muscle protein mass, function, and muscle quality also accompanies advancing age [60].

Hormonal changes are associated with a decline in lung function. A recent study suggests that alveolar hypoventilation and attendant arterial hypercapnia in healthy postmenopausal women are due in part to a reduction in circulating sex steroid hormone concentrations [61]. Longitudinal analysis of the relationship between basal plasma cortisol concentration and FEV_1 over an average of 4.7 years revealed a significant relationship between the plasma cortisol concentration and the rate of decline of FEV_1 after adjustment for age, height, smoking status, and initial FEV_1 [57]. Moreover, the multivariate model predicted that subjects with cortisol concentrations one standard deviation below the mean would experience FEV_1 declines of 71.6 mL/year greater than subjects with cortisol concentrations one standard deviation above the mean. The difference was comparable to an estimated 69.5 mL/year disparity between current smokers and never smokers [57]. These data indicate that physiological concentrations of cortisol may modulate the process responsible for the deterioration of ventilatory function with aging. A separate study of 631 male participants in the normative aging study (age range 44–85 years) showed that a two-hour urinary excretion of serotonin, but not 5-hydroxyindoleacetic acid (5-HIAA), decreased with age [62]. Animal studies suggest that the serotonin-dependent augmentation of respiratory motor output is reduced in old rats [63]. Current active smokers secrete significantly more serotonin than never smokers, while former smokers did not differ significantly from never smokers [62].

Hormonal changes can also significantly influence lung pathophysiology during early stages of lung growth. A recent study by Epaud and colleagues [64] found that the absence of IGF-1R significantly delayed the final stages of gestational lung maturation, although low levels of IGF-1R were sufficient to maintain normal lung development in mice. These findings indicate IGF-1R plays an essential role in cell proliferation and the timing of cell differentiation during fetal lung development while also possessing complex interactions with other factors, such as maternal corticosterone levels to impact on fetal lung maturation [65]. IGF-I/IGF-II is expressed in the human lung in early gestation, and the IGF family of growth factors, acting through the IGF-IR, appears to be essential to preserve endothelial cells during the process of neovascularization of the lung [66].

Since airway smooth muscle (ASM) tone is a primary regulator of airway hyperreactivity, changes in estrogen with age may play a significant role to prevent cholinergic-induced airway constriction via the activation of the nitric oxide–cGMP–protein kinase G pathway to increase calcium-activated potassium channel activity and indirectly reducing Ca^{2+} [67] to influence ASM contractility. Estrogen can also acutely and substantially reduce $[Ca^2]_i$ responses of human ASM cells to agonists such as histamine. These data support the idea of acute, nongenomic bronchodilatory effects of estrogens in the human airway [68], which could play a significant role in adult-onset asthma in life.

These age-related changes in the normal lung may predispose the elderly population to ventilatory failure during high demand states, such as with heart failure, pneumonia, or acute lung injury, with potentially fatal outcomes.

8.3   An animal model of the aging lung: The rat

Aging has been examined in a wide variety of animal species. Reviews of aging in mice, rats, and dogs have been previously reported [69, 70, 71]. From a practical experimental perspective, the rat provides a distinct number of advantages over most other laboratory species. These include ease in handling and maintaining over extended periods of time, a relatively short mean lifespan of 24–36 months of age, and an extensive literature base in chronic (2-year) toxicity studies (United States National Toxicology Program database: http://ntp-server.niehs.nih.gov/?objectid=72016020-BDB7-CEBA-F3E5A7965617C1C1). The two most commonly used rat strains for lifetime studies are Sprague Dawley and Fischer 344 rats. Lifespan characteristics of the Fischer 344 rat are well documented [72, 73, 74, 75, 76, 77, 78, 79, 80]. The majority of rats maintained under barrier conditions show no gross pathological changes of the lungs. Descriptions of age-related alterations in the respiratory system that do exist [76] have focused almost exclusively on the gas exchange portions of the lungs. Less information is available on detailed cellular and structural changes in the tracheobronchial airway tree with age with the exception of the postnatal changes in the nonciliated bronchiolar epithelial (Clara) cell [81, 82]. The majority of data related to lung aging in the rat is confined to brief descriptive observations of control animals in lifetime toxicity studies [83, 84].

8.3.1   The tracheobronchial tree and epithelium of the aging rat

The tracheobronchial epithelium of adult rats varies in terms of the types of cells present and the relative proportions of specific cell types throughout the conducting airway tree. In the trachea, four cell types have been identified: basal, serous, ciliated, and mucous goblet cells [85, 86]. In contrast, the epithelium of the bronchi and bronchioles consists of ciliated cells and nonciliated bronchiolar epithelial (Clara) cells [86, 87]. The composition of epithelial cells varies from proximal to distal airway generations with postnatal development [85, 87].

The transformation of tracheal epithelial cells in the rat beginning in the perinatal period through adult age appears to be continuous. At the time of birth, the rat trachea contains some mature ciliated cells, obvious secretary cells, and the beginning of basal cell differentiation [85, 88, 89]. Ciliogenesis begins at about 80% of the pregnancy time period in the rat.

Table 8.4 Volume density of epithelial cell of tracheobronchial airways of male Fischer 344 rats.

Epithelial cell type	Age (months)	Bronchus		
		Cranial	Central	Caudal
Ciliated cells	5	4.50 ± 0.88	2.74 ± 0.33	4.73 ± 0.33
	22	3.80 ± 0.56	3.33 ± 0.29	5.17 ± 0.73
Nonciliated cell	5	1.73 ± 0.30	2.07 ± 0.28	1.83 ± 0.04
	22	2.20 ± 0.52	1.89 ± 0.22	2.65 ± 0.45
Basal cells	5	0.00 ± 0.00	0.08 ± 0.03	0.02 ± 0.02
	22	0.00 ± 0.00	0.09 ± 0.06	0.00 ± 0.00
Total epithelial volume	5	6.23 ± 0.67	4.89 ± 0.43	6.58 ± 0.31
	22	6.01 ± 0.66	5.30 ± 0.55	7.82 ± 1.16

Values are presented as means ± SEM expressed as $\mu m^3/\mu m^2$; $n = 4$ for each age group.

Nonciliated secretary cells are obvious in the tracheal epithelium at about 90%–95% gestation.

A single study has evaluated the general characteristics of the airway epithelium in the aging rat. Pinkerton and colleagues [90] examined epithelial cells in three different airway generations – cranial (airway generation 5), central (airway generation 10), and caudal (airway generation 15) bronchi/bronchioles – in the left lung of the male Fischer 344 rats at 5 and 22 months of age. At each airway level, ciliated, nonciliated, and basal cells were identified, and the average volume density (cell volume/basal lamina surface area) of each cell type was determined (Table 8.4). No significant differences in the abundance of each cell type were noted from 5 to 22 months of age, and total epithelial cell volume remained constant during this time. Thus, the findings suggest little change in the proportion of epithelial cell types in the tracheobronchial tree with age.

8.3.2 Parenchymal lung structure in the aging rat

Total alveolar air space volume of the lungs progressively increases over a 2-year period of growth in Fischer 344 rats (Figure 8.1). Air space volume was measured in lungs intratracheally instilled with glutaraldehyde [76, 91]. Pulmonary physiology performed on male rats prior to glutaraldehyde instillation demonstrated the lungs were fixed at 75% of total lung capacity [76, 92, 93]. A marked increase in air space volume from 1 week to 5 months of age was associated with a significant increase in alveolar number. Randell and associates estimated the peak rate of formation for new alveoli in the neonatal rat to be 1000 per minute [94]. Light microscopic examination of large airways, terminal bronchioles, large blood vessels, alveoli, and alveolar septa demonstrated no detectable differences from 5 to 26 months of age. Animals 26 months of age appeared to have slightly enlarged alveolar ducts compared with those in younger age groups, but otherwise were indistinguishable from young animals. The percentage of the lungs constituting the lung parenchyma was 81% and 82% in young and old animals, respectively [76, 95].

Epithelial and capillary surface areas of the gas exchange regions of the lungs are given in Table 8.5. From 1 to 6 weeks of postnatal age, Fischer 344 rats had epithelial type I cell surface area increase sixfold and epithelial type II cell surface area increase threefold. Squamous type I cells covered more than 95% of the total alveolar surface and cuboidal type II cells covered the remaining 5%. The alveolar type III cell, a rare epithelial cell

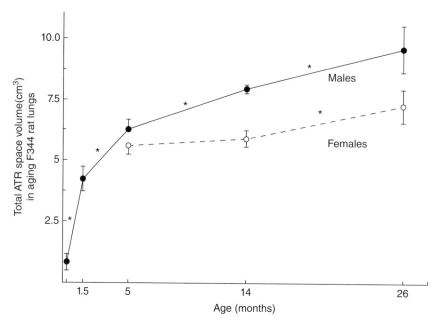

Figure 8.1 Total air space volume (cm³) in the glutaraldehyde-fixed lungs of male and female Fischer 344 rats. Each point represents the mean (±SD) of four animals. Asterisks indicate significant changes between consecutive age groups ($p < 0.05$). (Reproduced with permission from [69].)

found near bronchiole–alveolar duct junctions [87], contributed very little to the total alveolar surface area. The alveolar surface area of the Fischer 344 rats nearly doubled from 6 weeks to 5 months of age and then remained unchanged from 5 to 26 months of age. Since air space size increased by 50% from 5 to 26 months of age in the absence of any significant change in alveolar surface area, alveolar duct enlargement in older animals could explain, in part, how air space volume increased without a loss of surface area within the lung parenchyma. Enlargement of alveolar ducts was not quantitatively confirmed, but emphysematous changes (alveolar wall destruction) appeared to be absent in the aging Fischer 344 rats [75, 96, 97].

Changes in capillary surface area with age are proportional to changes in alveolar surface area (Table 8.5). Capillaries within the alveolar septa of the neonatal rat transform from a double capillary system to a single capillary system associated with the growth of secondary alveolar septa to form new alveoli by 3 weeks of age [76, 95]. From 5 to 26 months of age, the total surface area of the capillary bed in the lungs remains unchanged in both male and female rats (Table 8.5).

8.3.3 Alveolar tissue compartments

The morphometric characteristics of the major parenchymal cells making up the alveolar tissues over the lifespan of male and female Fischer 344 rats are presented in Table 8.6. The lung parenchyma consists of three tissue compartments: the epithelium, the interstitium, and the endothelium. Alveolar macrophages form a unique tissue compartment of individual cells that freely migrate along the surfaces of alveoli and airways and are also present within the pulmonary connective tissues as interstitial macrophages.

Table 8.5 Alveolar tissue volumes (cm^3/both lungs) in the aging Fischer 344 rats.

	1 week	6 weeks	5 months	14 months	26 months
Males					
No.	4	4	4	4	4
Total alveolar tissue	116 ± 14	296 ± 5*	427 ± 46*	454 ± 20	464 ± 43
Epithelium					
Type I	11 ± 2	61 ± 4*	82 ± 6*	84 ± 3	84 ± 8
Type II	7 ± 1	22 ± 3	37 ± 9	30 ± 4	28 ± 9
Interstitium					
Cellular	61 ± 6	77 ± 7	68 ± 5	57 ± 8	57 ± 7
Noncellular	13 ± 1	46 ± 3	128 ± 16*	151 ± 13	178 ± 29**
Endothelium	23 ± 4	84 ± 4*	94 ± 9	111 ± 6	96 ± 9
Macrophages	1 ± 0.4	6 ± 1	19 ± 7	21 ± 7	21 ± 7
Females					
No.	–	–	4	4	4
Total alveolar tissue	–	–	359 ± 15	313 ± 23	370 ± 22
Epithelium					
Type I	–	–	79 ± 7	58 ± 7*	66 ± 6**
Type II	–	–	25 ± 1	20 ± 3	18 ± 5
Interstitium	–	–			
Cellular	–	–	60 ± 7	49 ± 4	56 ± 6
Noncellular	–	–	81 ± 10	110 ± 11	153 ± 8**
Endothelium	–	–	93 ± 5	70 ± 4*	67 ± 5**
Macrophages	–	–	22 ± 9	6 ± 3	9 ± 3

Adapted from Pinkerton et al. (1982).
All values are mean ± SEM.
*$p < 0.05$ for comparison to the immediately younger group.
**$p < 0.05$ for comparison of the 26-month group to the 5-month group.

The composition of the alveolar epithelium in the Fischer 344 rat changes dramatically during postnatal development. The volume of type I epithelium increased more than sixfold and type II epithelium increased threefold from 1 to 6 weeks of age. The ratio of type II to type I cells in the lungs at 6 weeks of age was 1.8, but it decreased to 1.0 by 26 months of age [76]. The significance of the reduction in the type II to type I cell ratio is unknown; it may contribute to an altered secretory response in type II cells of old rats because a greater surface area must be served per type II cell to form the surfactant lining layer of the lungs compared with that in younger animals.

Over the lifespan of the Fischer 344 rat, the absolute volume of the cellular interstitium of the lung parenchyma does not change [98]. Interstitial cell number and cell size (Table 8.6) are similar at 1 week and 26 months of age, although the types and ratios of interstitial cell types forming the parenchyma cell pool (Table 8.6) are markedly different in neonatal pups compared to adults [99]. In contrast, the noncellular components of the interstitium demonstrate dramatic volume changes from 1 week through 5 months of age, increasing approximately tenfold in total volume [98]. From 5 to 26 months of age, interstitial volume continues to change at a lower but significant rate [98]. Compared to 5-month-old rats, the interstitial matrix increased 18% by 14 months of age and 39% by 26 months of age [98]. By 26 months of age, a further 89% increase in matrix volume was noted compared to 5 months of age [98]. These changes could be a result of pulmonary edema or an increase in the noncellular matrix components of the lung, such as collagen or elastin.

Table 8.6 Morphometric characteristics of the major parenchymal cells in the aging Fisher 344 rats.

	Males					Females		
	1 week	6 weeks	5 months	14 months	26 months	5 months	14 months	26 months
Alveolar type I cells								
Average volume (μm³)	709 ± 102	2,106 ± 121*	1,530 ± 121*	1,371 ± 151	1,462 ± 156	2,170	1,214 ± 125*	1,385 ± 120
Average surface area (μm²)	2,343 ± 145	8,608 ± 998	7612 ± 667	8,457 ± 1,122	8,699 ± 685	10,074 ± 1,615	7,431 ± 109	9,523 ± 863
Average basement membrane								
Surface area/cell (μm²)	2,147 ± 101	7,146 ± 679*	7,207 ± 755	7,478 ± 1,270	7,976 ± 702	9.146 ± 1.050	6,779 ± 163	8,528 ± 861
% total alveolar cell	6.7	6.5	8.1	11.4	11.2	8.4	11	10.3
% alveolar surface covered	91.3	96.1	96.4	98.6	98.6	97.6	98.3	98.6
Alveolar type II cells								
Average volume (μm³)	369 ± 60	401 ± 42	433 ± 108	326 ± 95	519 ± 178	341 ± 93	346 ± 27	332 ± 73
Average surface area (μm²)								
With microvilli	188 ± 20	292 ± 79	230 ± 41	223 ± 74	287 ± 218	347 ± 127	217 ± 73	217 ± 73
Without microvilli	183 ± 24	200 ± 55	185 ± 56	107 ± 33	124 ± 73	187 ± 64	97 ± 38	105 ± 38
% total alveolar cell	8.0	11.7	12.1	10.6	10.8	10.7	13	10.7
% alveolar surface covered	8.7	3.9	3.6	1.4	1.4	2.4	1.7	1.4
Interstitial cells								
Average volume (μm³)	513 ± 39	684 ± 67	427 ± 55*	452 ± 106	447 ± 51	456 ± 28	414 ± 40	411 ± 42
% total alveolar cell	52.7	24.8	24.4	24.2	24.7	28.5	27.5	28.8
Endothelial cells								
Average volume (μm³)	321 ± 44	334 ± 13	275 ± 25	394 ± 31	377 ± 55	421 ± 37	344 ± 28	303 ± 25
Average surface area (μm²)	459 ± 32	1,000 ± 32*	1,121 ± 95	1,618 ± 102*	1,725 ± 142*	1,467 ± 97	1,560 ± 198	1,604 ± 80
% total alveolar cell	31.7	55.0	51.0	51.3	49.9	47.9	46.9	46.8
Macrophages								
Average volume (μm³)	863 ± 143	672 ± 182	639 ± 131	1,488 ± 427	1,398 ± 513	1.091 ± 196	820 ± 444	551 ± 107
% total alveolar cell	0.9	2	4.3	2.5	3.2	4.3	1.8	3.6

Adapted from Pinkerton et al. (1982).
All values are mean ± SEM.
* *p* < 0.05 for comparison to the immediately younger group.
** *p* < 0.05 for comparison of the 26-month group to the 5-month group.

Vincent and colleagues [100] examined Fischer 344 rats for changes in the constituents of the interstitial matrix with age at the level of the proximal alveolar region (PAR), which consists of alveolar tissue sampled in a perpendicular orientation relative to the once axis of the terminal bronchiole approximately 300–400 μm down the alveolar ducts. Basement membrane thickness went from 40–45 nm at 4–6 months of age to 75–80 nm at 20–24 months. Similarly, collagen fiber volume, normalized to the surface of the alveolar epithelium, increased by more than 100% during this same period. Basement membrane volume and collagen fiber volume were similar in the PAR and accounted for 50% of the noncellular matrix at 4–6 months of age and 80% at 20–24 months of age. By comparison, no change was noted in the relative volume of elastin and remaining acellular space. The volume ratio of collagen fibers to elastin fibers shifted from 3 to 5 in the young adults (4–6 months) to 10 in the older animals (20–24 months). Mercer and Crapo [101] found the same collagen to elastin fiber ratio in Sprague Dawley rats as was seen in the young Fischer 344 rats.

The volume of ground substance relative to epithelial surface is not modified significantly in the lungs of the aging Fischer 344 rats. Therefore, it is unlikely that edema contributes significantly to the increase in interstitial matrix volume with age. The absence of detectable changes in the volume of elastin is consistent with its low rate of synthesis and slow turnover in adult animals (reviewed by [102, 103]). A progressive thickening of basement membranes may be explained by epithelial and endothelial cell turnover if the cells lay down basement membrane material during each cell cycle and possibly also during interphase. The increase in the volume of collagen fibers documented by morphometric approaches in adult rats confirms our previous morphologic observations [76] and is in agreement with biochemical studies of collagen done in the lungs of Lewis rats [104, 105] and Fischer 344 rats [106].

Morphologic and biochemical evidence supports the notion that there is a continuous deposition of mature cross-linked extracellular collagen in the lung parenchyma of aging rats, which can be interpreted as an age-related excess of collagen (or fibrosis). However, it should be noted that qualitative examination revealed no visible change in the overall gross architecture of the matrix in older animals compared to younger rats. Although speculative at present, it is conceivable that the dominant stimulus for collagen fiber deposition, through life, is the stress and strain exerted on fibers and translated to fibroblasts. Fibroblasts may add fibrils to the existing network while maintaining the general spatial relationship between collagen fibers and other tissue components and result in a net enrichment in collagen. It is not clear how this increase in collagen mass and the shift in the ratio of collagen to elastin with age impact the micromechanical behavior of the lungs. A logical assumption would be stiffer lungs, but physiological measurements suggest otherwise [76, 96]. Three-dimensional reconstructions in concert with physiological measurements [101, 107] would specifically address this issue in the senescent rat lung.

Mays et al. observed an abrupt fourfold decrease in the measured rate of gross collagen synthesis between 15 and 24 months of age in the Lewis rat [105]. More than 80% of newly synthesized collagen appears to be degraded intracellularly in 15-month-old animals, and 60% is degraded intracellularly at 24 months. The net result is only a small fraction of product being deposited as extracellular cross-linked collagen. Mays and associates suggest that maintenance of a high rate of collagen synthesis provides an adaptive capacity that allows the fibroblast to rapidly redirect procollagen from intracellular degradation pathways toward secretary pathways in response to injury or pathologic insult. There was no decline in fibroblast populations in aging lungs of Fischer 344 rat [76]. If the observations of Mays et al. [105] can be extended to other strains of rats, then a drop in the rate of collagen synthesis may constitute an intrinsic characteristic of the senescent lung fibroblast *in vivo* resulting in a compromised ability to repair collagen fibers.

The endothelium forms the third major tissue compartment of the alveolar septum. The volume of the endothelium in the lungs of Fischer 344 rats increased more than threefold from 1 to 6 weeks of age and then did not increase significantly through 26 months of age. The total surface area of the capillary endothelium, similar to the epithelial surface, demonstrated nearly a tenfold increase from 1 to 6 weeks of age and an additional 20% from 6 weeks to 5 months of age and did not change from 5 to 26 months of age. Endothelial cell size is relatively unchanged in male and female rats from 5 to 26 months of age (Table 8.6).

Alveolar macrophages form an important defense against inhaled particulates and pathogens in the lungs. Without these cells, the sterility of the lungs would be severely compromised. Alveolar macrophages at 1 week of age number approximately two million cells and increase to nine million cells by 6 weeks of age. Numbers are increased to 20 million cells by 5 months of age. Although no further increases in alveolar macrophage number are evident at 26 months of age [98], this population of cells is highly dynamic, and numbers can change rapidly through the recruitment of monocytes and macrophages from the interstitium and the blood [108] and by the *in situ* proliferation of cells within the lung air spaces [109].

8.4 Conclusion

The dynamics of lung growth, development, and aging are a continuous process, which involves every tissue compartment of the lungs. Because such changes may influence the response of the lungs to inhaled chemical agents and dusts, the age of any animal model should be considered in the evaluation of any experimental study. Different responses of the young versus aging lung to challenges may be due to changes within target cell populations and/or alterations in the functional status of cells through the aging process. Although cellular changes are most prominent during postnatal growth and development, modifications in cells continue through advanced age. Changes in cell number, size, and function associated with aging are likely to impact on lung physiology, metabolism, and immunity. Such changes could significantly alter the normal functions of the lung and its susceptibility to injury. Therefore, a more detailed understanding of the aging process is essential.

Acknowledgments

Background literature research, as presented here, has contributed to conceptual development of original research of the health effects of environmental air pollutants on the neonatal and aging lung as funded by US EPA Star Grant R831714, NIH ES00628, and NIH ES 116334. Underlying research is also supported in part by NIOSH grant 0H07550, NIEHS grant U01 ES 02027, RR 00169, and the California Tobacco-Related Disease Research Program (TRDRP) 18XT-0154. JYX was supported as Visiting Physician–Scientist Scholar from Dalian University Zhongshan Hospital, Dalian, China.

References

1. Anderson HR, Atkinson RW, Bremner SA, Marston L. Particulate air pollution and hospital admissions for cardiorespiratory diseases: Are the elderly at greater risk? *The European Respiratory Journal Supplement* 2003;40:39s–46s.
2. Kelly FJ. Vitamins and respiratory disease: Antioxidant micronutrients in pulmonary health and disease. *The Proceedings of the Nutrition Society* 2005;64:510–526.
3. Dyer C. The interaction of ageing and lung disease. *Chronic Respiratory Disease* 2012;9:63–67.
4. Ciencewicki J, Trivedi S, Kleeberger SR. Oxidants and the pathogenesis of lung diseases. *The Journal of Allergy and Clinical Immunology* 2008;122:456–468; quiz 469–470.
5. Janssens JP, Pache JC, Nicod LP. Physiological changes in respiratory function associated with ageing. *The European Respiratory Journal* 1999;13:197–205.
6. Zeleznik J. Normative aging of the respiratory system. *Clinics in Geriatric Medicine* 2003;19:1–18.
7. Guenard H, Marthan R. Pulmonary gas exchange in elderly subjects. *The European Respiratory Journal* 1996;9:2573–2577.
8. Sharma G, Goodwin J. Effect of aging on respiratory system physiology and immunology. *Clinical Interventions in Aging* 2006;1:253–260.
9. Peterson DD, Pack AI, Silage DA, Fishman AP. Effects of aging on ventilatory and occlusion pressure responses to hypoxia and hypercapnia. *The American Review of Respiratory Disease* 1981;124:387–391.
10. Pokorski M, Marczak M. Ventilatory response to hypoxia in elderly women. *Annals of Human Biology* 2003;30:53–64.
11. Garcia-Rio F, Villamor A, Gomez-Mendieta A, et al. The progressive effects of ageing on chemosensitivity in healthy subjects. *Respiratory Medicine* 2007;101:2192–2198.
12. Green F, Pinkerton K. Environmental determinants of lung aging. In: Harding R, Pinkerton K, Plopper C, editors. *The Lung: Development, Aging, and the Environment*. London, U.K.: Academic Press; 2004. pp. 377–395.
13. Mittman C, Edelman N, Norris A. Relationship between chest wall and pulmonary compliance with age. *Journal of Applied Physiology* 1965;20:1211–1216.
14. Polkey MI, Harris ML, Hughes PD, et al. The contractile properties of the elderly human diaphragm. *American Journal of Respiratory and Critical Care Medicine* 1997;155:1560–1564.
15. Gillooly M, Lamb D. Airspace size in lungs of lifelong non-smokers: Effect of age and sex. *Thorax* 1993;48:39–43.
16. D'Errico A, Scarani P, Colosimo E, Spina M, Grigioni WF, Mancini AM. Changes in the alveolar connective tissue of the ageing lung. An immunohistochemical study. *Virchows Archiv A, Pathological Anatomy and Histopathology* 1989;415:137–144.
17. Frette C, Jacob MP, Wei SM, et al. Relationship of serum elastin peptide level to single breath transfer factor for carbon monoxide in French coal miners. *Thorax* 1997;52:1045–1050.
18. Bennett WD, Zeman KL, Kim C. Variability of fine particle deposition in healthy adults: Effect of age and gender. *American Journal of Respiratory and Critical Care Medicine* 1996;153:1641–1647.
19. Azad N, Rojanasakul Y, Vallyathan V. Inflammation and lung cancer: Roles of reactive oxygen/nitrogen species. *Journal of Toxicology and Environmental Health Part B, Critical Reviews* 2008;11:1–15.
20. Kumagai Y, Koide S, Taguchi K, et al. Oxidation of proximal protein sulfhydryls by phenanthraquinone, a component of diesel exhaust particles. *Chemical Research in Toxicology* 2002;15:483–489.
21. Dick CA, Brown DM, Donaldson K, Stone V. The role of free radicals in the toxic and inflammatory effects of four different ultrafine particle types. *Inhalation Toxicology* 2003;15:39–52.
22. Davies KJ. The broad spectrum of responses to oxidants in proliferating cells: A new paradigm for oxidative stress. *IUBMB Life* 1999;48:41–47.
23. Abramova NE, Davies KJ, Crawford DR. Polynucleotide degradation during early stage response to oxidative stress is specific to mitochondria. *Free Radical Biology & Medicine* 2000;28:281–288.
24. Harman D. Ageing: Phenomena and theories. *Annals of New York Academic Science* 1998;854:1–7.
25. Rikans LE, Moore DR. Effect of aging on aqueous-phase antioxidants in tissues of male fischer rats. *Biochimica et Biophysica Acta* 1988;966:269–275.
26. Vincent R, Vu D, Hatch G, et al. Sensitivity of lungs of aging Fischer 344 rats to ozone: Assessment by bronchoalveolar lavage. *The American Journal of Physiology* 1996;271:L555–L565.

27. Bottje WG, Wang S, Beers KW, Cawthon D. Lung lining fluid antioxidants in male broilers: Age-related changes under thermoneutral and cold temperature conditions. *Poultry Science* 1998;77: 1905–1912.

28. Canada AT, Herman LA, Young SL. An age-related difference in hyperoxia lethality: Role of lung antioxidant defense mechanisms. *The American Journal of Physiology* 1995;268:L539–L545.

29. Meng Q, Wong YT, Chen J, Ruan R. Age-related changes in mitochondrial function and antioxidative enzyme activity in fischer 344 rats. *Mechanisms of Ageing and Development* 2007;128:286–292.

30. Cho HY, Reddy SP, Yamamoto M, Kleeberger SR. The transcription factor nrf2 protects against pulmonary fibrosis. *FASEB Journal* 2004;18:1258–1260.

31. Suzuki M, Betsuyaku T, Ito Y, et al. Down-regulated nf-e2-related factor 2 in pulmonary macrophages of aged smokers and patients with chronic obstructive pulmonary disease. *American Journal of Respiratory Cell and Molecular Biology* 2008;39:673–682.

32. Meyer KC. The role of immunity in susceptibility to respiratory infection in the aging lung. *Respiration Physiology* 2001;128:23–31.

33. Webster RG. Immunity to influenza in the elderly. *Vaccine* 2000;18:1686–1689.

34. Powers DC, Sears SD, Murphy BR, Thumar B, Clements ML. Systemic and local antibody responses in elderly subjects given live or inactivated influenza a virus vaccines. *Journal of Clinical Microbiology* 1989;27:2666–2671.

35. Kovaiou RD, Herndler-Brandstetter D, Grubeck-Loebenstein B. Age-related changes in immunity: Implications for vaccination in the elderly. *Expert Reviews in Molecular Medicine* 2007;9:1–17.

36. Yoshikawa TT. Important infections in elderly persons. *The Western Journal of Medicine* 1981;135: 441–445.

37. Litonjua AA, Sparrow D, Celedon JC, DeMolles D, Weiss ST. Association of body mass index with the development of methacholine airway hyperresponsiveness in men: The normative aging study. *Thorax* 2002;57:581–585.

38. Wang ML, McCabe L, Petsonk EL, Hankinson JL, Banks DE. Weight gain and longitudinal changes in lung function in steel workers. *Chest* 1997;111:1526–1532.

39. Gennuso J, Epstein LH, Paluch RA, Cerny F. The relationship between asthma and obesity in urban minority children and adolescents. *Archives of Pediatrics & Adolescent Medicine* 1998;152: 1197–1200.

40. Luder E, Melnik TA, DiMaio M. Association of being overweight with greater asthma symptoms in inner city black and hispanic children. *The Journal of Pediatrics* 1998;132:699–703.

41. Camargo CA, Jr., Weiss ST, Zhang S, Willett WC, Speizer FE. Prospective study of body mass index, weight change, and risk of adult-onset asthma in women. *Archives of Internal Medicine* 1999;159:2582–2588.

42. Guerra S, Sherrill DL, Bobadilla A, Martinez FD, Barbee RA. The relation of body mass index to asthma, chronic bronchitis, and emphysema. *Chest* 2002;122:1256–1263.

43. van den Borst B, Gosker HR, Koster A, et al. The influence of abdominal visceral fat on inflammatory pathways and mortality risk in obstructive lung disease. *The American Journal of Clinical Nutrition* 2012;96:516–526.

44. Womack CJ, Harris DL, Katzel LI, Hagberg JM, Bleecker ER, Goldberg AP. Weight loss, not aerobic exercise, improves pulmonary function in older obese men. *The Journals of Gerontology Series A, Biological Sciences and Medical Sciences* 2000;55:M453–M457.

45. Amara CE, Koval JJ, Paterson DH, Cunningham DA. Lung function in older humans: The contribution of body composition, physical activity and smoking. *Annals of Human Biology* 2001;28: 522–536.

46. Wills-Karp M. Age-related changes in pulmonary muscarinic receptor binding properties. *The American Journal of Physiology* 1993;265:L103–L109.

47. Hopp RJ, Bewtra A, Nair NM, Townley RG. The effect of age on methacholine response. *Journal of Allergy and Clinical Immunology* 1985;76:609–613.

48. Bennett BA, Mitzner W, Tankersley CG. The effects of age and carbon black on airway resistance in mice. *Inhalation Toxicology* 2012;24:931–938.

49. Connolly MJ, Crowley JJ, Charan NB, Nielson CP, Vestal RE. Impaired bronchodilator response to albuterol in healthy elderly men and women. *Chest* 1995;108:401–406.

50. Abrass IB, Scarpace PJ. Catalytic unit of adenylate cyclase: Reduced activity in aged-human lymphocytes. *The Journal of Clinical Endocrinology and Metabolism* 1982;55:1026–1028.

51. Creticos P, Knobil K, Edwards LD, Rickard KA, Dorinsky P. Loss of response to treatment with leukotriene receptor antagonists but not inhaled corticosteroids in patients over 50 years of age. *Annals of Allergy, Asthma & Immunology* 2002;88:401–409.

52. Korenblat PE, Kemp JP, Scherger JE, Minkwitz MC, Mezzanotte W. Effect of age on response to zafirlukast in patients with asthma in the accolate clinical experience and pharmacoepidemiology trial (accept). *Annals of Allergy, Asthma & Immunology* 2000;84:217–225.

53. Grubor B, Ramirez-Romero R, Gallup JM, Bailey TB, Ackermann MR. Distribution of substance p receptor (neurokinin-1 receptor) in normal ovine lung and during the progression of bronchopneumonia in sheep. *The Journal of Histochemistry and Cytochemistry* 2004;52:123–130.

54. Ferrari E, Cravello L, Muzzoni B, et al. Age-related changes of the hypothalamic-pituitary-adrenal axis: Pathophysiological correlates. *European Journal of Endocrinology/European Federation of Endocrine Societies* 2001;144:319–329.

55. Nass R, Thorner MO. Impact of the gh-cortisol ratio on the age-dependent changes in body composition. *Growth Hormone & IGF Research* 2002;12:147–161.

56. Fanciulli G, Delitala A, Delitala G. Growth hormone, menopause and ageing: No definite evidence for 'rejuvenation' with growth hormone. *Human Reproduction Update* 2009;15:341–358.

57. Sparrow D, O'Connor GT, Rosner B, DeMolles D, Weiss ST. A longitudinal study of plasma cortisol concentration and pulmonary function decline in men. The normative aging study. *The American Review of Respiratory Disease* 1993;147:1345–1348.

58. Matousek RH, Sherwin BB. Sex steroid hormones and cognitive functioning in healthy, older men. *Hormones and Behavior* 2010;57:352–359.

59. Tuohimaa P, Keisala T, Minasyan A, Cachat J, Kalueff A. Vitamin d, nervous system and aging. *Psychoneuroendocrinology* 2009;34 Suppl 1:S278–S286.

60. Walston J, Hadley EC, Ferrucci L, et al. Research agenda for frailty in older adults: Toward a better understanding of physiology and etiology: Summary from the American geriatrics society/national institute on aging research conference on frailty in older adults. *Journal of the American Geriatrics Society* 2006;54:991–1001.

61. Preston ME, Jensen D, Janssen I, Fisher JT. Effect of menopause on the chemical control of breathing and its relationship with acid-base status. *American Journal of Physiology Regulatory, Integrative and Comparative Physiology* 2009;296:R722–R727.

62. Sparrow D, O'Connor GT, Young JB, Rosner B, Weiss ST. Relationship of urinary serotonin excretion to cigarette smoking and respiratory symptoms. The normative aging study. *Chest* 1992;101:976–980.

63. Zabka AG, Behan M, Mitchell GS. Long term facilitation of respiratory motor output decreases with age in male rats. *The Journal of Physiology* 2001;531:509–514.

64. Epaud R, Aubey F, Xu J, et al. Knockout of insulin-like growth factor-1 receptor impairs distal lung morphogenesis. *PLoS One* 2012;7:e48071.

65. Silva D, Venihaki M, Guo WH, Lopez MF. Igf2 deficiency results in delayed lung development at the end of gestation. *Endocrinology* 2006;147:5584–5591.

66. Han RN, Post M, Tanswell AK, Lye SJ. Insulin-like growth factor-i receptor-mediated vasculogenesis/angiogenesis in human lung development. *American Journal of Respiratory Cell and Molecular Biology* 2003;28:159–169.

67. Dimitropoulou C, White RE, Ownby DR, Catravas JD. Estrogen reduces carbachol-induced constriction of asthmatic airways by stimulating large-conductance voltage and calcium-dependent potassium channels. *American Journal of Respiratory Cell and Molecular Biology* 2005;32:239–247.

68. Townsend EA, Thompson MA, Pabelick CM, Prakash YS. Rapid effects of estrogen on intracellular Ca^{2+} regulation in human airway smooth muscle. *American Journal of Physiology Lung Cellular and Molecular Physiology* 2010;298:L521–L530.

69. Pinkerton KE, Vincent R, Plopper CG, Young SL. Normal development growth and aging of the lung. In: Mohr U, Dungworth DL, Capen CC, editors. *Pathobiology of the Aging Rat*. Washington, DC: ILSI Press; 1992. pp. 97–109.

70. Pinkerton KE, Cowin LL, Witschi H. Development growth and aging of the lungs. In: Mohr U, Dungworth DL, Capen CC, Carlton WW, Sundberg JP, Ward JM, editors. *Pathobiology of the Aging Mouse*. Washington, DC: ILSI Press; 1996. pp. 261–272.

71. Pinkerton KE, Murphy KM, Hyde DM. Morphology and morphometry of the lung. In: Capen CC, Benjamin SA, Hahn FF, Dungworth DL, Carlton WW, editors. *Pathobiology of the Aging Dog*. Ames, IA: Iowa State University Press; 2001. pp. 43–56.

72. Boorman G, Eustis S. Lung. In: Boorman G, Eustis S, Elwell M, Montgomery C, MacKenzie W, editors. *Pathology of the Fischer Rat*. New York: Academic Press; 1990. pp. 339–367.

73. Chesky J, Rockstein M. Lifespan characteristics in the male fischer rat. *Experimental Aging Research* 1976;2:399–407.

74. Jacob BB, Huseby RA. Neoplasms occurring in aged fischer rats with special reference to testicular, uterine, and thyroid tumors. *Journal of National Cancer Institute* 1967;39:303–309.

75. Mauderly JL, Likens SA. Relationships of age and sex to function of fischer 344 rats. *Federation Proceedings* 1980;39:10–91.

76. Pinkerton KE, Barry BE, O'Neil JJ, Raub JA, Pratt PC, Crapo JD. Morphologic changes in the lung during the lifespan of fischer 344 rats. *The American Journal of Anatomy* 1982;164:155–174.

77. Rockstein M, Chesky J, Sussman M. Comparative biology and evolution of aging. In: Finch C, Hayflick L, editors. *Handbook of the Biology of Aging*. New York: Van Nostrand Reinhold; 1977. pp. 3–34.

78. Snell K. Spontaneous lesions of the rat. In: Ribelin W, McCoy J, editors. *The Pathology of Laboratory Animals*. Springfield, IL: Thomas; 1965. pp. 211–302.

79. Sass B, Rabstein L, Madison R, Nims R, Peters R, Kelloff G. Incidence of spontaneous neoplasms in f344 rats throughout the natural life-span. *Journal of National Cancer Institute* 1975;54:1449–1456.

80. Masoro EJ. Mortality and growth characteristics of rat strains commonly used in aging research. *Experimental Aging Research* 1980;6:219–233.

81. Massaro GD, Davis L, Massaro D. Postnatal development of the bronchiolar Clara cell in rats. *The American Journal of Physiology* 1984;247:C197–C203.

82. Massaro GD, Massaro D. Development of bronchiolar epithelium in rats. *The American Journal of Physiology* 1986;250:R783–R788.

83. Coleman GL, Barthold W, Osbaldiston GW, Foster SJ, Jonas AM. Pathological changes during aging in barrier-reared fischer 344 male rats. *Journal of Gerontology* 1977;32:258–278.

84. Goodman DG, Ward JM, Squire RA, Chu KC, Linhart MS. Neoplastic and nonneoplastic lesions in aging f344 rats. *Toxicology and Applied Pharmacology* 1979;48:237–248.

85. Jeffery P, Reid L. The ultrastructure of the airway lining and its development. In: Hodson W, editor. *The Development of the Lung*. New York: Marcel Dekker Inc.; 1977. pp. 87–134.

86. Plopper CG. Comparative morphologic features of bronchiolar epithelial cells. The Clara cell. *The American Review of Respiratory Disease* 1983;128:S37–S41.

87. Chang LY, Mercer RR, Crapo JD. Differential distribution of brush cells in the rat lung. *The Anatomical Record* 1986;216:49–54.

88. Cirelli E. Elektronenmikroskopische analyze der pra und postnatalen differen-zierung des epithels der oberen luftwege der ratte. *Zeitschrift für mikroskopisch-anattomische Forschung* 1966;41:132–178.

89. Kober H. The luminal surface of the rat trachea during ontogenesis [in German]. *Zeitschrift für mikroskopisch-anatomische Forschung* 1975;89:99–409.

90. Pinkerton KE, Weller BL, Menache MG, Plopper CG. Consequences of prolonged inhalation of ozone on f344/n rats: Collaborative studies. Part xiii. A comparison of changes in the tracheobronchial epithelium and pulmonary acinus in male rats at 3 and 20 months. *Research Report (Health Effects Institute)* 1998:1–32; discussion 33–37.

91. Crapo JD, Peters-Golden M, Marsh-Salin J, Shelburne JS. Pathologic changes in the lungs of oxygen-adapted rats: A morphometric analysis. *Laboratory Investigation* 1978;39:640–653.

92. Hayatdavoudi G, Crapo JD, Miller FJ, O'Neil JJ. Factors determining degree of inflation in intratracheally fixed rat lungs. *Journal of Applied Physiology* 1980;48:389–393.

93. Takezawa J, Miller FJ, O'Neil JJ. Single-breath diffusing capacity and lung volumes in small laboratory mammals. *Journal of Applied Physiology* 1980;48:1052–1059.

94. Randell SH, Mercer RR, Young SL. Postnatal growth of pulmonary acini and alveoli in normal and oxygen-exposed rats studied by serial section reconstructions. *The American Journal of Anatomy* 1989;186:55–68.

95. Burri P. The postnatal growth of the rat lung, iii: Morphology. *The Anatomical Record* 1974;180:77–98.

96. Mauderly JL. Effect of age on pulmonary structure and function of immature and adult animals and man. *Federation Proceedings* 1979;38:173–177.

97. Liebow A. Summary: Biochemical and structural changes in the aging lung. In: Cander L, Moyer J, editors. *Aging of the Lung*. New York: Grune and Stratton; 1964. pp. 97–104.

(a)

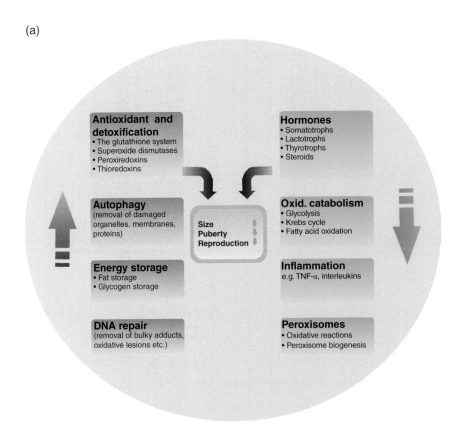

Plate 2.1 Components and instigators of the survival response in mammals. (a) The survival response represents a preservative metabolic strategy that aims at removing free radicals, toxins, damaged organelles, and membranes; delays age-related genomic decay and in the meantime suppresses growth and reproduction; and decreases inflammatory responses. In mammals, the various components of this conserved response result in smaller body size, delayed puberty, and decreased reproduction.

Molecular Aspects of Aging: Understanding Lung Aging, First Edition. Edited by Mauricio Rojas, Silke Meiners and Claude Jourdan LeSaux.
© 2014 John Wiley & Sons, Inc. Published 2014 by John Wiley & Sons, Inc.

(b)

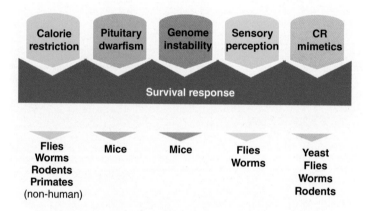

Plate 2.1 (b) Shown are the major instigators that are presently known to trigger a generalized evolutionary survival strategy against a multitude of adverse physiological threats. However, there might be more cues involved in triggering the battery of physiological responses favoring longevity.

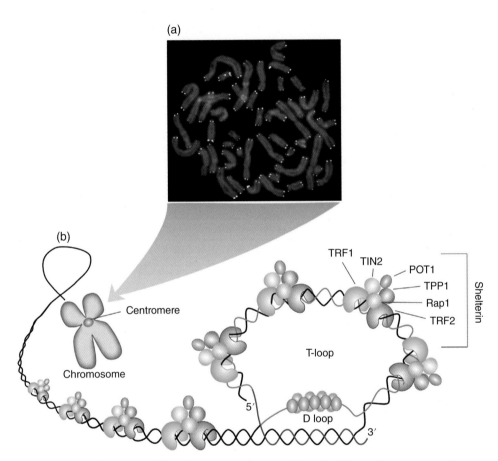

Plate 4.1 (a) Telomeres are ribonucleoprotein structures at the ends of linear chromosomes and can be detected by fluorescence *in situ* hybridization (FISH, yellow bright signals) and the extremities of each chromatide of the chromosomes (blue structures) during metaphase. (b) Schematically, telomeres are composed of hundreds to thousands of TTAGGG hexameric DNA repeats coated by specialized proteins collectively termed shelterin, forming a lariat at the very end of the molecule (T-loop).

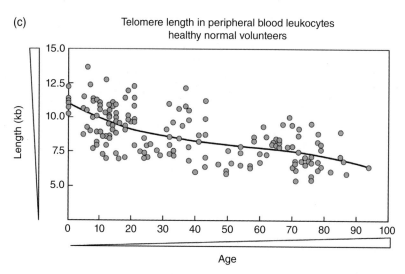

(c)

Telomere length in peripheral blood leukocytes
healthy normal volunteers

Plate 4.1 (c) Telomeres shorten with human aging. The graphic represents the lengths of telomeres of peripheral blood leukocytes from birth to 100 years in healthy volunteers. Telomeres are measured in kilobases (kb) and each yellow circle represents one individual.

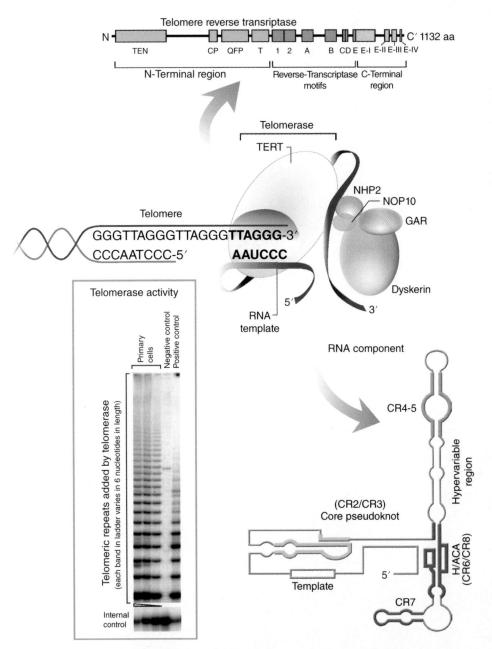

Plate 4.2 The telomerase holoenzyme is composed of the catalytic unit, TERT, its RNA component (TERC) that serves as a template for telomere elongation, and associated proteins (dyskerin, GAR, NOP10, and NHP2). The complex attaches to the 3′ end of the telomeric DNA and adds hexameric repeats. The inset shows a method to evaluate telomerase activity. Telomerase enzymatic activity may be measured *in vitro* by PCR amplification of telomerase products that are run in a gel. Each band represents the telomerase product and is separated by six nucleotides.

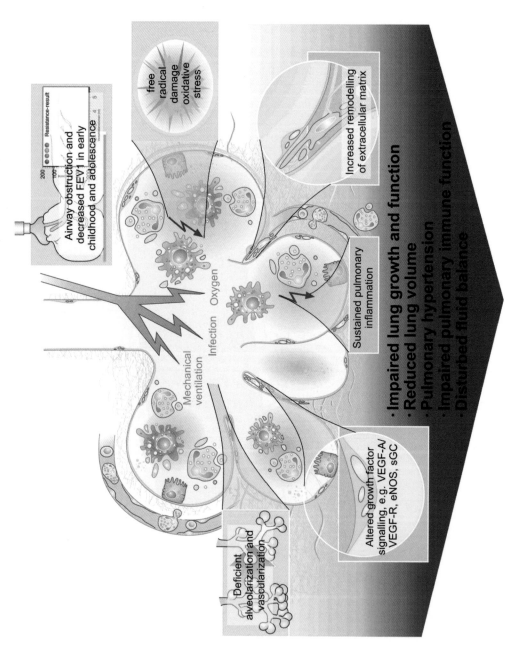

Plate 10.1 Summary of mechanisms that mediate characteristic changes to the neonatal lung following injury and a brief overview of the possible consequences.

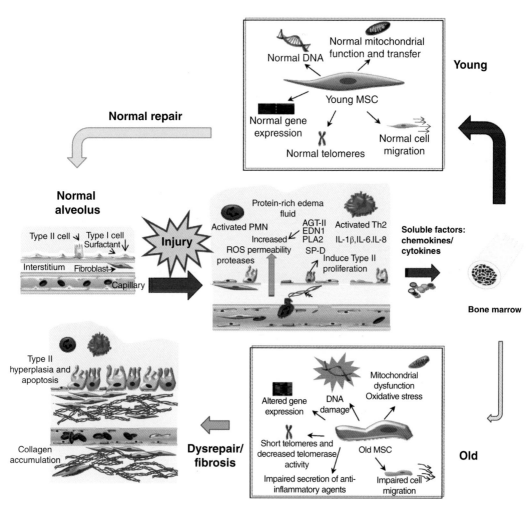

Plate 12.1 Comparison of the mechanisms of response in young and aged individuals. In the normal alveolus (left), there is normal fluid movement from the vascular to the interstitial space, with normal architecture of the epithelium and the endothelium, and production of surfactant. The panel in the middle shows an injured alveolus. Injury activates macrophages, which secrete a battery of proinflammatory cytokines and chemokines that leak into the bloodstream. Among these cytokines, IL-8 is responsible for the recruitment of neutrophils into the injured alveolus. ROS and proteases released by the activated neutrophils damage the alveolar epithelium and endothelium, and as a result, there is an increase in the permeability and in the influx of protein-rich edema. Another characteristic during injury is the increased levels of surfactant protein D by type II alveolar epithelial cells. The top right panel shows the representation of a restored alveolus including recruitment of healthy MSCs. MSCs themselves produce and stimulate the production of IL-10 by macrophages, which attenuates their inflammatory cascade. The secretion of angiopoietin 1 and the transfer of mitochondria to the epithelial cells by MSCs together with the production of KGF contribute to restoring the epithelium and endothelium. This results in improved alveolar fluid clearance and decreased permeability. The bottom right panel shows a defective senescent MSC with a decrease in their ability to respond to activation and mobilization, resulting in lung disrepair and fibrosis.

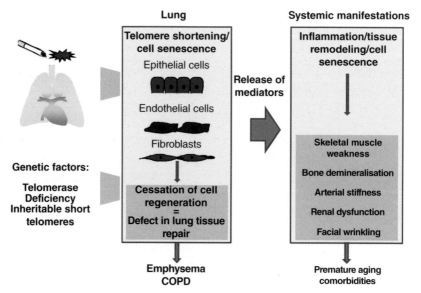

Plate 13.1 Diagram illustrating potential mechanisms linking cell senescence and development of lung alterations at the origin of COPD and systemic manifestations of the disease.

98. Pinkerton KE, Green FHY. Normal aging of the lung. In: Harding R, Pinkerton KE, Plopper CG, editors. *The Lung: Development, Aging, and the Environment*. London, U.K.: Elsevier Academic Press; 2004. pp. 213–233.

99. Brody JS, Kaplan NB. Proliferation of alveolar interstitial cells during postnatal lung growth. Evidence for two distinct populations of pulmonary fibroblasts. *The American Review of Respiratory Disease* 1983;127:763–770.

100. Vincent R, Mercer RR, Chang LY. Morphometric study of interstitial matrix in the lungs of the aging rat. *FASEB Journal* 1990;4:A1915.

101. Mercer RR, Crapo JD. Spatial distribution of collagen and elastin fibers in the lungs. *Journal of Applied Physiology* 1990;69:756–765.

102. Foster JA, Curtiss SW. The regulation of lung elastin synthesis. *The American Journal of Physiology* 1990;259:L13–L23.

103. Rucker RB, Dubick MA. Elastin metabolism and chemistry: Potential roles in lung development and structure. *Environmental Health Perspectives* 1984;55:179–191.

104. Mays PK, Bishop JE, Laurent GJ. Age-related changes in the proportion of types i and iii collagen. *Mechanisms of Ageing and Development* 1988;45:203–212.

105. Mays PK, McAnulty RJ, Laurent GJ. Age-related changes in lung collagen metabolism. A role for degradation in regulating lung collagen production. *The American Review of Respiratory Disease* 1989;140:410–416.

106. Sahebjami H. Lung tissue elasticity during the lifespan of fischer 344 rats. *Experimental Lung Research* 1991;17:887–902.

107. Mercer RR, Crapo JD. Three-dimensional reconstruction of the rat acinus. *Journal of Applied Physiology* 1987;63:785–794.

108. Brain JD, Sorokin S, Godieski IJ. Quantification, origin, and fate of pulmonary macrophages. In: Brain JD, Proctor DF, Reid LM, editors. *Lung Biology in Health and Disease: Respiratory Defense Mechanisms*. New York: Dekker; 1977. p. 849.

109. Shellito J, Esparza C, Armstrong C. Maintenance of the normal rat alveolar macrophage cell population. The roles of monocyte influx and alveolar macrophage proliferation in situ. *The American Review of Respiratory Disease* 1987;135:78–82.

9 Mouse Models to Explore the Aging Lung

Mingyi Wang[1] and Deepak A. Deshpande[2]

[1] Intramural Research Program, National Institute on Aging, Baltimore, Maryland, USA
[2] Pulmonary and Critical Care Medicine Division, University of Maryland School of Medicine, Baltimore, Maryland, USA

Aging is a multidimensional process resulting in decreasing homeostatic balance and organ system functions that contribute to diseases or diminished quality of life. Respiration is one important organ system function affected. Aging is associated with increased work of breathing and increased incidences of lung diseases. Aging involves progressive functional impairment, which is often associated with reduced capacity to respond to stress and injury. This declining ability leads to homeostatic imbalance and organ system malfunction resulting in increased risks for chronic obstructive pulmonary disease (COPD) and idiopathic pulmonary fibrosis (IPF) with advanced age. Respiration is regulated at multiple levels through a complex, coordinated collaboration of different cells and organ systems. Similarly, diseases of pulmonary system involve interaction among resident airway cells, pathogens, and immune cells. Understanding molecular and cellular basis of respiratory changes due to age is critical in clinical management of respiratory failure and other age-associated pulmonary ailments. Mouse models provide unique advantages of being able to study molecular pathogenesis in the face of multiple organ systems and aging.

9.1 Pulmonary changes during aging

Fundamental changes with age in lung mechanics are well established, and these changes can be further aggravated by diseases affecting the lung [1]. In elderly subjects, clinical studies measuring various lung volumes and capacities (spirometry) reveal that total lung capacity (TLC) remains unchanged but vital capacity (VC), forced expiratory volume in 1 s (FEV1), and forced vital capacity (FVC) decline rapidly and an increase in residual volume (RV) forces aged subjects to breathe at a higher lung volume [1, 2, 3, 4]. In addition, peak flow rate is lower in the elderly with a characteristic decline in maximal inspiratory flow. These changes not only affect various aspects of respiration but also the ability to deliver inhaled drugs. Despite our appreciation of the age-dependent decline in respiratory function, the basis for this decline is poorly understood. Establishing a good animal model is the first step

Molecular Aspects of Aging: Understanding Lung Aging, First Edition. Edited by Mauricio Rojas, Silke Meiners and Claude Jourdan Le Saux.
© 2014 John Wiley & Sons, Inc. Published 2014 by John Wiley & Sons, Inc.

in discerning cellular and molecular basis for the age-dependent changes in respiratory functions. The ability to determine respiratory functions *in vivo* and *ex vivo* using whole animal or isolated tissues (trachea, lungs, lung slices) in mice has made mouse a suitable model in pulmonary biomedical, genetic, pathological, pharmacological, and toxicological research. Similarly, a murine model is of great interest because of the prominent role played by mice in studies of diseases affecting the lungs such as asthma, COPD, sepsis, fibrosis, and lung cancer. The prominence of these conditions in the older population has been well established [5, 6, 7, 8]. Relevance of the mouse model in respiratory studies to pulmonary functions and diseases in humans has been debated for a long time [9, 10, 11]. Nevertheless, mouse studies have offered great insights into molecular pathogenesis of respiratory diseases and identified new therapeutic targets. A good animal model should be able to recapitulate most, if not all, prominent clinical features observed in humans. In that respect, the mouse model is suitable for pulmonary aging studies.

Bronchoprovocative responsiveness (as measured by change in forced expiratory flow at 50% VC, i.e., FEF50) to albuterol, a primary therapeutic agent used in the treatment of asthma, was reported to be lower in older healthy subjects compared to adults suggesting β-adrenergic function changes with age [12, 13, 14, 15, 16]. Upon methacholine challenge, the increase in airway resistance was found to be diminished in the elderly, as analysis of pressure–volume relationship demonstrated a leftward and upward shift compared to that observed for the healthy adults. Methacholine reactivity is significantly compromised in the elderly and confirmed in animal models [16, 17, 18, 19]. Noninvasive measurements of inspiratory and expiratory pressures also reveal an age-dependent decline in respiratory function. Mouse models are very effective in replicating bronchoprovocative responses to adrenergic and cholinergic agonists seen in humans. Molecular and cellular basis for age-dependent changes in airway responsiveness can be further studied using mouse models.

9.1.1 Advantages of mouse models for studying physiological lung changes

Aging is a multifactorial complex physiological process. Mice and humans share a number of anatomical and functional characteristics. System, organ, cellular, and gene structures are similar in mice compared to humans [20, 21, 22]. Furthermore, organ and systemic physiology (including lung function) of mice share common features with humans. Most importantly, molecular entities that regulate cell differentiation, development, cellular/ organ functions, and senescence or death in mice are similar to those described in humans. In many instances, disease pathogenesis (including lung diseases) and cellular changes in response to etiological factors are similar in mice and humans. The mechanisms of cellular senescence and aging are similar in mice compared to humans although several differences also exist (e.g., rates of production of reactive oxygen species and metabolism are different in mice compared to humans) [20, 21, 22]. Nevertheless, due to vast similarities between mice and humans, the mouse model of aging can be exploited to understand human aging [20, 21, 22]. Mouse models have the following advantages to study age-dependent changes in lung function:

- Mice can be maintained in a controlled environment without confounding effects of environmental factors influencing aging in the lungs. Hence, mice provide a good model for determining effects of age *per se*.

- Relatively short time needed to study age effects.
- Low cost of maintenance.
- Well-characterized genetic information.
- Ability to manipulate gene expression – transgenics, knockouts, knock-ins – can be generated in both global and tissue-specific manner.
- Inbred strains are readily available.
- Lung functions can be measured by invasive and noninvasive methods. Noninvasive measurements are ideal for longitudinal studies to test the effect of aging or any drug treatments to manipulate *in vivo* lung functions.
- Histology of normal and pathological lungs is well characterized.
- The immune system is well established – useful to determine immunosenescence and immunological basis for pulmonary diseases affecting humans due to age.
- Alveolar fluid analysis is a good indicator of lung pathologies.
- Tissues can be harvested for studies *ex vivo* (trachea, lung slices, whole lung) and *in vitro*.
- Airway cells and immune cells migrated into the lungs can be isolated and studied *in vitro* individually.
- Dietary modifications (e.g., calorie restriction) can be easily and accurately performed.
- Aerosol treatment can be performed to prevent systemic effects of drugs.

Mice can be used in pulmonary function studies by employing invasive and noninvasive techniques. Noninvasive protocols are ideal for aging studies as they can be performed in live animals and suitable for longitudinal studies in the same animal. The procedure does not require anesthesia and animals will be breathing naturally. The limitations of this technique include indirect measurements of pulmonary function and poor quality of data due to measurement artifacts [9, 23]. Invasive measurements on the other hand are very accurate, directly measure pulmonary functions, exclude contribution of upper airways, and hence are very well suited for measuring changes in lung functions due to obstructive airway diseases. Although the procedure is technically challenging, the data directly reflects pulmonary changes. Major limitations include requirement of tracheal/orotracheal intubation, anesthesia, and euthanasia at the end of the measurements [24]. Depending upon the design and objectives of the study, one may choose either invasive or noninvasive measurements to determine age effect on lung functions.

9.2 Key findings from mouse models of aging

Hirai and colleagues studied age-dependent changes in lung functions using accelerated senescence-resistant strain of mice, SAMR-1 [25]. Lung capacity, dynamic compliance, and static compliance were found to increase depending upon the age of the mice [25, 26]. The investigators concluded that the increases in lung volume primarily determine changes in lung mechanics due to age. In two different studies, Huang et al. measured pressure–volume curves and lung impedance in different age groups of mice [27, 28]. The authors compared age effects in two different strains (C57BL/6 and DBA/2J) of mice. Not so surprisingly, the authors reported differences in mechanics between the two strains of mice as well as age effect. However, lung volume significantly increased with age (measured as lung volume at 30 cmH$_2$O (V30) pressure). Age-dependent increase in lung volume was associated with decreased elastic recoil pressure and airway resistance (measured as positive end-expiratory pressure) [27, 28].

Human and animal model studies have revealed decreased elastin and increased collagen content in the lungs. Similarly, studies using animal models have established that aging results in a loss of lung elastance and a leftward shift in the pressure–volume curve [27, 28]. However, in mouse models, it was observed that the pattern of structural changes does not coincide with changes in lung mechanics. Other studies in mouse models have focused on identifying global gene expression changes in the lungs [29, 30] that correlate to the functional changes in lung mechanics. Microarray analysis of RNA isolated from the lungs of different age groups revealed up- and down-regulation of several important genes. Downregulated genes include extracellular matrix genes such as collagen, Adamts2, Fbn1, matrix metalloprotease (Mmp)-2, Mmp14, and Eln. These genes are critically important in the maintenance of extracellular matrix, elasticity, and integrity of alveolar structures in the lung. Similarly, many genes in the lungs were upregulated depending upon the age. Observations from such studies are very important in understanding structural basis for loss of functions in the lungs. However, future studies are needed to delineate global gene expression patterns in the specific cell types.

9.2.1 Longevity and lung function in mice

Studies from lower vertebrates and transgenic animals have identified a few genes that determine longevity. It is not clear if modifications in the expression of longevity genes help prevent age-dependent changes in lung functions or onset of pathologies. Few examples of mouse models are described later that require further exploration in the context of pulmonary functions.

Candidate gene and genome-wide association studies (GWAS) for longevity comparison of human and mouse genomes revealed an important finding [31, 32]. A series of quantitative trait loci identified on human genome were associated with longevity. Interestingly, eight of the ten top human associations were located within a previously reported mouse life-span quantitative trait loci. The findings demonstrate that many genes or pathways leading to age-associated changes in humans are shared by mouse. Therefore, the mouse model may help speed the understanding of the genetic or genomic basis of biology of aging.

9.2.2 Different strains of mice have different alterations in lung mechanics

Different strains of mice are known to have different median survival age. In one study, the authors explored the differences in lung functions in DBA/2J strain of mice that has a shorter life span (21 months) and C57BL/6 strain (life span of 29 months) [28]. These two stains have also shown variations in lung pathology during emphysema (cigarette smoke induced), ovalbumin-induced eosinophilia, and pneumonitis [33, 34, 35]. Loss of lung functions during aging was widely different in DBA/2J strain compared to that in C57BL/6 strain of mice [28]. Hence, selection of age groups for aging studies requires careful selection of mice strain. Microarray studies on gene expression profiles in mice lungs revealed a greater number of upregulated genes in the lungs of DBA/2J mice compared with C57BL/6 mice with age, suggesting a genomic basis for accelerated aging in DBA/2J mice [29]. Information on strain differences are very significant in pulmonary function and lung disease studies and will come to aid, enabling researchers to take quick and correct decisions while selecting appropriate animal models. Lung function differences among different strains of mice under both normal and after ovalbumin sensitization and challenge have been well established [35, 36] and further demonstrate the genetic basis for lung functions and diseases. Bleomycin-induced lung injury model is the most commonly used mouse model to study IPF. The fibrotic response to bleomycin in mice is strain dependent. C57BL/6 mice are highly susceptible to bleomycin-induced fibrosis, whereas DBA/2 mice are intermediate, and the Balb/c mice are less susceptible [37]. This strain-dependent variation reflects differences in the expression of the inactivating enzyme, bleomycin hydrolase. Most importantly, not all the inbred strains of mice survive for an extended period of time [38, 39]. This is further underscored by the fact that bleomycin-induced lung fibrotic response in old (24 months) mice was found to be different than the response observed in young (3 months) mice [40]. Hence, age groups to be included in the study depend upon the strain of the mice used.

9.2.3 Transgenic mouse model to study aging in the lungs

A list of genetically modified mice that could be used to establish age-dependent cellular and molecular changes and to study pathogenesis of age-associated pulmonary diseases is provided in Table 9.1.

9.2.3.1 Adenylyl cyclase (AC) 5 knockout mice

Mouse lacking adenylyl cyclase (AC) 5, an enzyme that catalyzes the conversion of ATP into cyclic adenosine monophosphate (cAMP), has been shown to be protected from oxidative stress and hence live longer (~30% more) [72, 52]. The protective effect of deleting AC5 expression was well studied in cardiac and skeletal tissues. However, AC/cAMP/protein kinase A pathway has been shown to play a critical role in the regulation of airway functions under normal and inflammatory conditions [73, 74]. A detailed evaluation of pulmonary changes in the AC5 knockout mice may provide new insights into molecular basis for pulmonary aging.

9.2.3.2 Klotho knockout mice

klotho gene encodes a transmembrane protein that acts as coreceptor for fibroblast growth factor (FGF). Loss of *klotho* gene function results in premature aging, and multiple organs show structural and functional changes similar to normal aging [75, 49]. *klotho*-deficient mice are well suited for aging studies. At 6–8 weeks of age, $klotho^{-/-}$ mice develop significant lung pathology and functional changes such as increase in lung volume, dynamic compliance, and mean expiration time and decrease in tidal volume [76, 77, 78]. Corresponding structural changes are characterized by enlargement of air space and increased destruction index similar to emphysema. Overall, *klotho* knockout mouse may represent a good model to study senile lung.

9.2.3.3 Senescence-accelerated mouse (SAM: Senescence prone (SAMP) and senescence resistant (SAMR))

Inbred mice lines of AKR/J strain were developed that are either senescence prone (SAMP) or senescence resistant (SAMR). The 14 different SAMP lines manifest varying degrees of senescence-associated changes in different organ structure and function. The lung changes

Table 9.1 List of genetically modified mice that could be used to establish age-dependent cellular and molecular changes and pathogenesis of age-associated pulmonary diseases.

Transgenic		Knockout		Mutations		
Longevity	Ref.	Longevity	Ref.	Longevity		Ref.
PEPCK-c (phosphoenolpyruvate carboxykinase, cytosolic)	[41]	p66shc (growth factor adaptor Shc-66 kDa isoform)	[42]	Ames dwarf	Prop-1	[43]
αMUPA (urokinase-type plasminogen activator)	[44, 45]	PAPPA (pregnancy-associated plasma protein A)	[46]	Snell dwarf	Pit-1	[47, 48]
Klotho	[49]	Irs-1 (insulin receptor substrate)	[50]	Little	GHRH	[51]
Oxidative stress		AC5 (adenylyl cyclase)	[52]	Laron dwarf	GH receptor	[53]
Metallothionein	[54]	Surf1 (COX assembly factor)	[55]			
Mitochondrial catalase	[56]	S6K1 (ribosomal protein S6 kinase)	[57]			
		Artr1a (angiotensin II type 1 receptor)	[58]			
Metabolic		RIIb (PKA subunit)	[94]			
IGF-1 (insulin-like growth factor)	[54]	Klotho	[46]			
		GH	[54, 59]			
		Oxidative stress				
		Nrf (nuclear factor erythroid 2-related factor)	[95, 96]			
		SOD (superoxide dismutase)	[97]			
		EC matrix				
		Relaxin	[98, 99]			
		SMP-30 (senescence marker protein)	[62]			
		Cav-1 (caveolin-1)	[64]			
		Decorin	[65]			
		Metabolic				
		IGF-1 (insulin-like growth factor)	[54]			
		Mclk1 (mitochondrial CLK-1)	[100]			
		FGF-23 (fibroblast growth factor)	[101]			

observed with age are similar to human senile lung. Size of air spaces and change in pressure–volume curves show rapid increase with age in SAMP mice compared to SAMR mice [79, 80, 81, 82]. Therefore, SAM mouse can be used as a good model to study pulmonary aging.

9.2.3.4 Growth hormone (GH) and growth hormone receptor (GHR)

The role of growth hormone (GH) and growth hormone receptor (GHR) in aging has been studied extensively using mouse models. Interestingly, GH knockout mice live shorter with increased incidences of age-associated diseases, whereas GHR knockout mice live longer [60–85]. Proteomic analysis of the lungs obtained from GH and GHR knockout mouse revealed the critical role of GH in early lung growth, oxidative protection, lipid and energy metabolism, and proteasomal activity [61, 86, 87]. Detailed investigations of pulmonary changes and development of respiratory diseases in these animals may reveal new pathways involved in pulmonary aging.

9.2.3.5 Senescence marker protein (SMP)-30

Senescence marker protein (SMP)-30 is a 30 kDa protein expressed in a variety of cell types. Expression of SMP-30 decreases with age resulting in an increased apoptosis of cells. Pulmonary changes have been briefly studied in SMP-30 knockout mice with particular emphasis on emphysema [67, 88, 89]. Future *in vitro* and *in vivo* studies using SMP-30 mice may generate novel pathways involved in age-dependent changes in the lungs.

9.2.3.6 Caveolin-1

Caveolae are plasma membrane invaginations containing structural proteins such as Caveolin-1 (Cav-1) and several other important signaling elements including receptors, enzymes, and cytoskeletal elements. Using Cav-1 knockout mice, studies have shown the role of Cav-1 in age-related pulmonary diseases such as emphysema and fibrosis [68].

9.2.3.7 Decorin

Synthesis of collagen, a major component of lung parenchyma, requires decorin. Structural and functional evaluation of the lungs in decorin knockout mice revealed loss of development of normal lung parenchyma and decreased airway resistance and compliance [69]. This study suggests that decorin is critical in the development of the lung. What role decorin plays in the development of age-associated changes in lung structure and function needs further evaluation.

Table 9.1 lists genetically modified (overexpression, knockout, and mutant genes) mice that could provide insights into age-associated changes in lung structure and functions. These mice could also be used to establish molecular pathogenesis of age-dependent pulmonary diseases.

9.3 Age is a risk factor for obstructive pulmonary diseases

In addition to change in pulmonary functions with age, the incidence of pulmonary obstructive diseases also increases with age. The common chronic respiratory diseases diagnosed in the elderly include COPD, emphysema, and IPF [6, 88, 90, 91]. Environmental factors such as smoking, toxins, pollutants, ozone, and infections trigger the development and progression of age-associated pulmonary diseases. Very interestingly, the pathogenesis of COPD, emphysema, and IPF shares largely cellular and molecular changes reported in aging. For example, oxidative stress, immunosenescence, decline in repair capacity, change in telomere length, and altered intracellular signaling and gene expression are commonly seen in emphysema, COPD, and aging. IPF is a widely studied pulmonary disease, and more importantly, molecular pathogenesis of IPF has been largely derived from animal model studies [92, 93]. The mouse model of IPF is arguably the most commonly used model. Bleomycin, silica/asbestos, irradiation, gene overexpression, or fluorescein isothiocyanate is used to induce IPF in mice (reviewed in [94]). Each model has its own limitations and advantages. Nevertheless, bleomycin-induced pulmonary fibrosis is most widely used mouse model of IPF. These models are used not only to study pathogenesis but also to test various therapeutic agents to treat IPF. For example, bleomycin model has been used to test antifibrotic activity of doxycycline, erythromycin, ONO-1301, and IMD-0354 (reviewed in [92]). Age is one of the important criteria while selecting mice for studying lung diseases as aged lungs show increased susceptibility to injury and development of fibrosis compared to the lungs from young animals. For example, bleomycin-induced lung fibrosis is more severe in the lungs from old mice compared to young mice [40]. Similarly, other agents such as cigarette smoking, LPS, and mechanical ventilation that induce lung injury have shown increased severity in older mice compared to young mice [95, 96, 97, 98]. Hence, age of the animals needs to be considered while designing experiments using mouse models.

COPD is another pulmonary disease of old age. Etiology of COPD is not well understood; however, smoking is an important risk factor for the development of this disease. Mouse models of COPD are well established and widely used to study the pathogenesis of COPD [68, 90]. Other environmental factors such as pollutants and ozone are also used to study pulmonary effects using mouse models. Cigarette smoke exposure models have come a long way in improving purity of the tobacco products, precision of exposure time, concentration, and frequency. Mouse models of second-hand smoking are also established. Interestingly, both IPF and COPD share several features of pathogenesis although clinical manifestation of both the diseases is very different [6]. Premature **senescence** of the resident airway cells is believed to be the primary mechanism that underlies development of IPF and COPD (reviewed in [6]). Environmental factors such as smoking cause oxidative stress in the pulmonary cells and induce a series of changes that cause premature senescence of the cells [90, 91]. Therefore, understanding the molecular basis of age-associated changes in lung structure and functions is pivotal for elucidating the pathogenesis of obstructive pulmonary diseases and developing effective therapies. Definitely, mouse models play a critical role in these scientific endeavors.

9.4 Challenges ahead

Extrapolation to human aging studies from mouse model in a very controlled environment poses challenges as human respiratory effects are significantly influenced by the constant exposure to environmental factors. Incorporating environmental factors in aging studies using murine model is almost obligatory in many experiments [99, 100]. Separating age effects *per se* from other comorbidities such as other systemic diseases that occur during old age is an important factor to consider while interpreting the data from murine model. Strain variation in the phenotype adds another layer of complexity to the data interpretation.

Table 9.2 Factors to be considered while selecting mice for studying aging lung.

Criteria	Good practice	Comments
Age groups	3 or more groups	
	Not too old and not too young	Median survival age for the strain is a good parameter
Strain of mice	Hybrid strains are appropriate	Inbred stains are genetically homogeneous
	Repeat studies in two different stains	
	Use same strain for all the studies	Airway responses are strain dependent
	Transgenic mice – use at least 3 different lines	
Number of animals	Keep extra mice in each groups	Aged animals tend to die
	Do not use a minimum number of animals	
	Avoid pooling of animal tissues/cells/samples	Between-animal variation is significantly high in aging
Disease model	Select model appropriate for the phenotype	Pulmonary disease phenotype is protocol dependent
	Strain of the animal	Strain-specific variations in disease severity and phenotype are very common
Statistics	Use appropriate analysis	Important to differentiate age and disease effects on outcomes

Identifying specific targets to improve or maintain normal lung functions is challenging although genetic studies from mouse model provide molecular basis of age-dependent changes in lung structure/function. Recent advances in inducible cell-type-specific knockout strategies may help in overcoming the inherent difficulties of data interpretation of global transgenic animal models. Differences in the design of the experiments and methodology used to study age and disease effects on pulmonary functions significantly influence the experimental outcomes. Therefore, establishing standardized guidelines to study age and age-associated pulmonary diseases is valuable. In this context, general guidelines for morphometric and quantitative analysis of lung structures have been developed [101]. The use of standardized guidelines may help in minimizing technical errors in interpreting experimental data. Some important criteria to be considered while selecting the mouse model to study aging are listed in Table 9.2. Finally, mathematical modeling of data from mouse models using advanced algorithms would be ideal to extrapolate mouse data to humans. The proposed models have several limitations and need extensive improvisation.

9.5 Conclusion

The aging lung can be considered from several different perspectives. These include alterations in pulmonary functions, therapeutic responses in aging lungs, structural basis of age-dependent changes in lung functions, and cellular and molecular changes in pulmonary cells such as the epithelium, smooth muscle, fibroblast, and connective tissue. The concept of animal models is a prerequisite for translational research. A successful animal model has to reproduce the characteristic features of the physiological changes and diseases affecting the human lung in a way so as to improve our understanding of the mechanisms involved. A number of organisms have been used as models to study aging with each one having strengths and weaknesses in different aspects. Mice have proved to be the most widely used animal model as they are cheap and easy to maintain and bear resemblances to the human organism and there are genetic tools that allow manipulations not possible in other organisms. More importantly, mice allow us to test hypotheses generated from human studies and provide *in vivo* relevance for observations from basic science studies. Considering vast applications of mice in physiological and pathological aging experiments, scientists should be able to utilize mouse models to study pulmonary aging and establish molecular basis for the respiratory changes in the elderly.

Acknowledgments

This work was supported by the National Institute on Aging grant (AG041265) to Deshpande.

References

1. Janssens JP, Pache JC, Nicod LP. Physiological changes in respiratory function associated with ageing. *Eur Respir J* 1999;13:197–205.
2. Brandstetter RD, Kazemi H. Aging and the respiratory system. *Med Clin North Am* 1983;67:419–431.
3. Ericsson P, Irnell L. Effect of five years' ageing on ventilatory capacity and physical work capacity in elderly people. *Acta Med Scand* 1969;185:193–199.
4. Ericsson P, Irnell L. Spirometric studies of ventilatory capacity in elderly people. *Acta Ned Scand* 1969;185:179–184.
5. Blagosklonny MV. Prevention of cancer by inhibiting aging. *Cancer Biol Ther* 2008;7:1520–1524.
6. Chilosi M, Poletti V, Rossi A. The pathogenesis of COPD and IPF: distinct horns of the same devil? *Respir Res* 2012;13:3.
7. De Gaudio AR, Rinaldi S, Chelazzi C, Borracci T. Pathophysiology of sepsis in the elderly: clinical impact and therapeutic considerations. *Curr Drug Targets* 2009;10:60–70.
8. Jones SC, Iverson D, Burns P, Evers U, Caputi P, Morgan S. Asthma and ageing: an end user's perspective – the perception and problems with the management of asthma in the elderly. *Clin Exp Allergy* 2011;41:471–481.
9. Bates JH, Irvin CG. Measuring lung function in mice: the phenotyping uncertainty principle. *J Appl Physiol* 2003;94:1297–1306.
10. Bates JH, Rincon M, Irvin CG. Animal models of asthma. *Am J Physiol Lung Cell Mol Physiol* 2009;297:L401–L410.
11. Irvin CG, Bates JH. Measuring the lung function in the mouse: the challenge of size. *Respir Res* 2003;4:4.
12. Connolly MJ. Ageing, late-onset asthma and the beta-adrenoceptor. *Pharmacol Ther* 1993;60:389–404.
13. Connolly MJ, Crowley JJ, Charan NB, Nielson CP, Vestal RE. Impaired bronchodilator response to albuterol in healthy elderly men and women. *Chest* 1995;108:401–406.
14. Connolly MJ, Shaw L. Respiratory disease in old age: research into Ageing Workshop, London, 1998. *Age Ageing* 2000;29:281–285.
15. Duncan PG, Brink C, Douglas JS. Beta-receptors during aging in respiratory tissues. *Eur J Pharmacol* 1982;78:45–52.
16. Hankinson JL, Odencrantz JR, Fedan KB. Spirometric reference values from a sample of the general U.S. population. *Am J Respir Crit Care Med* 1999;159:179–187.
17. Hopp RJ, Bewtra A, Nair NM, Townley RG. The effect of age on methacholine response. *J Allergy Clin Immunol* 1985;76:609–613.
18. Wills-Karp M. Age-related changes in pulmonary muscarinic receptor binding properties. *Am J Physiol* 1993;265:L103–L109.
19. Wills M, Douglas JS. Aging and cholinergic responses in bovine trachealis muscle. *Br J Pharmacol* 1988;93:918–924.
20. Demetrius L. Of mice and men. When it comes to studying ageing and the means to slow it down, mice are not just small humans. *EMBO Rep* 2005;6 Spec No.:S39–S44.

21. Demetrius L. Aging in mouse and human systems: a comparative study. *Ann N Y Acad Sci* 2006;1067:66–82.
22. Yuan R, Peters LL, Paigen B. Mice as a mammalian model for research on the genetics of aging. *ILAR J* 2011;52:4–15.
23. Lundblad LK, Irvin CG, Hantos Z, Sly P, Mitzner W, Bates JH. Penh is not a measure of airway resistance! *Eur Respir J* 2007;30:805.
24. Glaab T, Taube C, Braun A, Mitzner W. Invasive and noninvasive methods for studying pulmonary function in mice. *Respir Res* 2007;8:63.
25. Hirai T, Hosokawa M, Kawakami K, et al. Age-related changes in the static and dynamic mechanical properties of mouse lungs. *Respir Physiol* 1995;102:195–203.
26. Takubo Y, Hirai T, Muro S, Kogishi K, Hosokawa M, Mishima M. Age-associated changes in elastin and collagen content and the proportion of types I and III collagen in the lungs of mice. *Exp Gerontol* 1999;34:353–364.
27. Huang K, Rabold R, Schofield B, Mitzner W, Tankersley CG. Age-dependent changes of airway and lung parenchyma in C57BL/6J mice. *J Appl Physiol* 2007;102:200–206.
28. Huang K, Mitzner W, Rabold R, et al. Variation in senescent-dependent lung changes in inbred mouse strains. *J Appl Physiol* 2007;102:1632–1639.
29. Misra V, Lee H, Singh A, et al. Global expression profiles from C57BL/6J and DBA/2J mouse lungs to determine aging-related genes. *Physiol Genomics* 2007;31:429–440.
30. Tankersley CG, Shank JA, Flanders SE, et al. Changes in lung permeability and lung mechanics accompany homeostatic instability in senescent mice. *J Appl Physiol* 2003;95:1681–1687.
31. Murabito JM, Yuan R, Lunetta KL. The search for longevity and healthy aging genes: insights from epidemiological studies and samples of long-lived individuals. *J Gerontol A Biol Sci Med Sci* 2012;67:470–479.
32. Walter S, Atzmon G, Demerath EW, et al. A genome-wide association study of aging. *Neurobiol Aging* 2011;32:2109–2128.
33. Bartalesi B, Cavarra E, Fineschi S, et al. Different lung responses to cigarette smoke in two strains of mice sensitive to oxidants. *Eur Respir J* 2005;25:15–22.
34. Gudmundsson G, Monick MM, Hunninghake GW. IL-12 modulates expression of hypersensitivity pneumonitis. *J Immunol* 1998;161:991–999.
35. Whitehead GS, Walker JK, Berman KG, Foster WM, Schwartz DA. Allergen-induced airway disease is mouse strain dependent. *Am J Physiol Lung Cell Mol Physiol* 2003;285:L32–L42.
36. Irvin CG. Using the mouse to model asthma: the cup is half full and then some. *Clin Exp Allergy* 2008;38:701–703.
37. Schrier DJ, Kunkel RG, Phan SH. The role of strain variation in murine bleomycin-induced pulmonary fibrosis. *Am Rev Respir Dis* 1983;127:63–66.
38. Sundberg JP, Berndt A, Sundberg BA, et al. The mouse as a model for understanding chronic diseases of aging: the histopathologic basis of aging in inbred mice. *Pathobiol Aging Age Relat Dis* 2011;1.
39. Yuan R, Tsaih SW, Petkova SB, et al. Aging in inbred strains of mice: study design and interim report on median lifespans and circulating IGF1 levels. *Aging Cell* 2009;8:277–287.
40. Sueblinvong V, Neujahr DC, Mills ST, et al. Predisposition for disrepair in the aged lung. *Am J Med Sci* 2012;344:41–51.
41. McGrane MM, de VJ, Yun J, et al. Tissue-specific expression and dietary regulation of a chimeric phosphoenolpyruvate carboxykinase/bovine growth hormone gene in transgenic mice. *J Biol Chem* 1988;263:11443–11451.
42. Migliaccio E, Giorgio M, Mele S, et al. The p66shc adaptor protein controls oxidative stress response and life span in mammals. *Nature* 1999;402:309–313.
43. Brown-Borg HM, Bartke A. GH and IGF1: roles in energy metabolism of long-living GH mutant mice. *J Gerontol A Biol Sci Med Sci* 2012;67:652–660.
44. Miskin R, Masos T, Yahav S, Shinder D, Globerson A. AlphaMUPA mice: a transgenic model for increased life span. *Neurobiol Aging* 1999;20:555–564.
45. Miskin R, Tirosh O, Pardo M, et al. AlphaMUPA mice: a transgenic model for longevity induced by caloric restriction. *Mech Ageing Dev* 2005;126:255–261.
46. Conover CA, Bale LK. Loss of pregnancy-associated plasma protein A extends lifespan in mice. *Aging Cell* 2007;6:727–729.
47. Flurkey K, Papaconstantinou J, Miller RA, Harrison DE. Lifespan extension and delayed immune and collagen aging in mutant mice with defects in growth hormone production. *Proc Natl Acad Sci U S A* 2001;98:6736–6741.

48. Flurkey K, Papaconstantinou J, Harrison DE. The Snell dwarf mutation Pit1(dw) can increase life span in mice. *Mech Ageing Dev* 2002;123:121–130.
49. Kuro-o M. Klotho and aging. *Biochim Biophys Acta* 2009;1790:1049–1058.
50. Selman C, Lingard S, Choudhury AI, et al. Evidence for lifespan extension and delayed age-related biomarkers in insulin receptor substrate 1 null mice. *FASEB J* 2008;22:807–818.
51. Eicher EM, Beamer WG. Inherited ateliotic dwarfism in mice. Characteristics of the mutation, little, on chromosome 6. *J Hered* 1976;67:87–91.
52. Yan L, Vatner DE, O'Connor JP, et al. Type 5 adenylyl cyclase disruption increases longevity and protects against stress. *Cell* 2007;130:247–258.
53. Zhou Y, Xu BC, Maheshwari HG, et al. A mammalian model for Laron syndrome produced by targeted disruption of the mouse growth hormone receptor/binding protein gene (the Laron mouse). *Proc Natl Acad Sci U S A* 1997;94:13215–13220.
54. Yang X, Doser TA, Fang CX, et al. Metallothionein prolongs survival and antagonizes senescence-associated cardiomyocyte diastolic dysfunction: role of oxidative stress. *FASEB J* 2006;20:1024–1026.
55. Dell'agnello C, Leo S, Agostino A, et al. Increased longevity and refractoriness to Ca(2+)-dependent neurodegeneration in Surf1 knockout mice. *Hum Mol Genet* 2007;16:431–444.
56. Schriner SE, Linford NJ. Extension of mouse lifespan by overexpression of catalase. *Age (Dordr)* 2006;28:209–218.
57. Selman C, Tullet JM, Wieser D, et al. Ribosomal protein S6 kinase 1 signaling regulates mammalian life span. *Science* 2009;326:140–144.
58. Benigni A, Corna D, Zoja C, et al. Disruption of the Ang II type 1 receptor promotes longevity in mice. *J Clin Invest* 2009;119:524–530.
59. Enns LC, Pettan-Brewer C, Ladiges W. Protein kinase A is a target for aging and the aging heart. *Aging (Albany NY)* 2010;2:238–243.
60. Berryman DE, Christiansen JS, Johannsson G, Thorner MO, Kopchick JJ. Role of the GH/IGF-1 axis in lifespan and healthspan: lessons from animal models. *Growth Horm IGF Res* 2008;18:455–471.
61. Beyea JA, Sawicki G, Olson DM, List E, Kopchick JJ, Harvey S. Growth hormone (GH) receptor knockout mice reveal actions of GH in lung development. *Proteomics* 2006;6:341–348.
62. Iizuka T, Ishii Y, Itoh K, et al. Nrf2-deficient mice are highly susceptible to cigarette smoke-induced emphysema. *Genes Cells* 2005;10:1113–1125.
63. Ishii Y, Itoh K, Morishima Y, et al. Transcription factor Nrf2 plays a pivotal role in protection against elastase-induced pulmonary inflammation and emphysema. *J Immunol* 2005;175:6968–6975.
64. Sentman ML, Brannstrom T, Marklund SL. EC-SOD and the response to inflammatory reactions and aging in mouse lung. *Free Radic Biol Med* 2002;32:975–981.
65. Samuel CS, Zhao C, Bathgate RA, et al. Relaxin deficiency in mice is associated with an age-related progression of pulmonary fibrosis. *FASEB J* 2003;17:121–123.
66. Samuel CS, Zhao C, Bathgate RA, et al. The relaxin gene-knockout mouse: a model of progressive fibrosis. *Ann N Y Acad Sci* 2005;1041:173–181.
67. Mori T, Ishigami A, Seyama K, et al. Senescence marker protein-30 knockout mouse as a novel murine model of senile lung. *Pathol Int* 2004;54:167–173.
68. Zou H, Stoppani E, Volonte D, Galbiati F. Caveolin-1, cellular senescence and age-related diseases. *Mech Ageing Dev* 2011;132:533–542.
69. Fust A, LeBellego F, Iozzo RV, Roughley PJ, Ludwig MS. Alterations in lung mechanics in decorin-deficient mice. *Am J Physiol Lung Cell Mol Physiol* 2005;288:L159–L166.
70. Liu X, Jiang N, Hughes B, Bigras E, Shoubridge E, Hekimi S. Evolutionary conservation of the clk-1-dependent mechanism of longevity: loss of mclk1 increases cellular fitness and lifespan in mice. *Genes Dev* 2005;19:2424–2434.
71. DeLuca S, Sitara D, Kang K, et al. Amelioration of the premature ageing-like features of Fgf-23 knockout mice by genetically restoring the systemic actions of FGF-23. *J Pathol* 2008;216:345–355.
72. Hu CL, Chandra R, Ge H, et al. Adenylyl cyclase type 5 protein expression during cardiac development and stress. *Am J Physiol Heart Circ Physiol* 2009;297:H1776–H1782.
73. Billington CK, Penn RB. Signaling and regulation of G protein-coupled receptors in airway smooth muscle. *Respir Res* 2003;4:2.
74. Deshpande DA, Penn RB. Targeting G protein-coupled receptor signaling in asthma. *Cell Signal* 2006;18:2105–2120.
75. Kuro-o M. Disease model: human aging. *Trends Mol Med* 2001;7:179–181.
76. Ishii M, Yamaguchi Y, Yamamoto H, Hanaoka Y, Ouchi Y. Airspace enlargement with airway cell apoptosis in klotho mice: a model of aging lung. *J Gerontol A Biol Sci Med Sci* 2008;63:1289–1298.

77. Nabeshima Y. Klotho: a fundamental regulator of aging. *Ageing Res Rev* 2002;1:627–638.
78. Suga T, Kurabayashi M, Sando Y, et al. Disruption of the klotho gene causes pulmonary emphysema in mice. Defect in maintenance of pulmonary integrity during postnatal life. *Am J Respir Cell Mol Biol* 2000;22:26–33.
79. Carter TA, Greenhall JA, Yoshida S, et al. Mechanisms of aging in senescence-accelerated mice. *Genome Biol* 2005;6:R48.
80. Teramoto S, Fukuchi Y, Uejima Y, Teramoto K, Oka T, Orimo H. A novel model of senile lung: senescence-accelerated mouse (SAM). *Am J Respir Crit Care Med* 1994;150:238–244.
81. Teramoto S, Fukuchi Y, Uejima Y, Teramoto K, Orimo H. Biochemical characteristics of lungs in senescence-accelerated mouse (SAM). *Eur Respir J* 1995;8:450–456.
82. Uejima Y, Fukuchi Y, Nagase T, Tabata R, Orimo H. A new murine model of aging lung: the senescence accelerated mouse (SAM)-P. *Mech Ageing Dev* 1991;61:223–236.
83. Coschigano KT, Holland AN, Riders ME, List EO, Flyvbjerg A, Kopchick JJ. Deletion, but not antagonism, of the mouse growth hormone receptor results in severely decreased body weights, insulin, and insulin-like growth factor I levels and increased life span. *Endocrinology* 2003;144:3799–3810.
84. Ding J, Sackmann-Sala L, Kopchick JJ. Mouse models of growth hormone action and aging: a proteomic perspective. *Proteomics* 2013;13:674–685.
85. Masternak MM, Al-Regaiey KA, Del Rosario Lim MM, et al. Caloric restriction and growth hormone receptor knockout: effects on expression of genes involved in insulin action in the heart. *Exp Gerontol* 2006;41:417–429.
86. Beyea JA, Olson DM, Harvey S. Growth hormone (GH) action in the developing lung: changes in lung proteins after adenoviral GH overexpression. *Dev Dyn* 2005;234:404–412.
87. Beyea JA, Olson DM, Harvey S. Growth hormone-dependent changes in the rat lung proteome during alveorization. *Mol Cell Biochem* 2009;321:197–204.
88. Fukuchi Y. The aging lung and chronic obstructive pulmonary disease: similarity and difference. *Proc Am Thorac Soc* 2009;6:570–572.
89. Sato T, Seyama K, Sato Y, et al. Senescence marker protein-30 protects mice lungs from oxidative stress, aging, and smoking. *Am J Respir Crit Care Med* 2006;174:530–537.
90. Tuder RM, Petrache I. Pathogenesis of chronic obstructive pulmonary disease. *J Clin Invest* 2012;122:2749–2755.
91. Tuder RM, Kern JA, Miller YE. Senescence in chronic obstructive pulmonary disease. *Proc Am Thorac Soc* 2012;9:62–63.
92. Moeller A, Ask K, Warburton D, Gauldie J, Kolb M. The bleomycin animal model: a useful tool to investigate treatment options for idiopathic pulmonary fibrosis? *Int J Biochem Cell Biol* 2008;40:362–382.
93. Mouratis MA, Aidinis V. Modeling pulmonary fibrosis with bleomycin. *Curr Opin Pulm Med* 2011;17:355–361.
94. Moore BB, Hogaboam CM. Murine models of pulmonary fibrosis. *Am J Physiol Lung Cell Mol Physiol* 2008;294:L152–L160.
95. Lin SP, Sun XF, Chen XM, Shi SZ, Hong Q, Lv Y. Effect of aging on pulmonary ICAM-1 and MCP-1 expressions in rats with lipopolysaccharide-induced acute lung injury. *Nan Fang Yi Ke Da Xue Xue Bao* 2010;30:584–587.
96. Matulionis DH. Chronic cigarette smoke inhalation and aging in mice: 1. Morphologic and functional lung abnormalities. *Exp Lung Res* 1984;7:237–256.
97. Moriyama C, Betsuyaku T, Ito Y, et al. Aging enhances susceptibility to cigarette smoke-induced inflammation through bronchiolar chemokines. *Am J Respir Cell Mol Biol* 2010;42:304–311.
98. Nin N, Lorente JA, de PM, et al. Aging increases the susceptibility to injurious mechanical ventilation. *Intensive Care Med* 2008;34:923–931.
99. Bennett BA, Mitzner W, Tankersley CG. The effects of age and carbon black on airway resistance in mice. *Inhal Toxicol* 2012;24:931–938.
100. Sunil VR, Patel KJ, Mainelis G, et al. Pulmonary effects of inhaled diesel exhaust in aged mice. *Toxicol Appl Pharmacol* 2009;241:283–293.
101. Hsia CC, Hyde DM, Ochs M, Weibel ER. How to measure lung structure – what for? On the "Standards for the quantitative assessment of lung structure". *Respir Physiol Neurobiol* 2010;171:72–74.

10 Evidence for Premature Lung Aging of the Injured Neonatal Lung as Exemplified by Bronchopulmonary Dysplasia

Anne Hilgendorff

Comprehensive Pneumology Center (CPC), University Hospital, Ludwig-Maximilians University, Helmholtz Zentrum München; Member of the German Center for Lung Research (DZL); Dr. von Haunersches Children's Hospital Munich, Germany

10.1 Introducing bronchopulmonary dysplasia

Chronic lung disease of the newborn, also known as bronchopulmonary dysplasia (BPD), is the most common chronic lung disease in early infancy and results in an increased risk for pulmonary and neurologic impairment persisting into adulthood [1]. The disease is defined by the need for supplemental oxygen and/or ventilator support for greater than 28 days, or beyond 36 weeks postmenstrual age (PMA), and thereby classified into three different severity grades (mild, moderate, severe) [1].

The incidence of BPD varies between newborn care centers, reflecting differences in patient population and infant management practices [2, 3, 4, 5]. Recent publications report an incidence of BPD of up to 68% in very low birth weight (401–1500 g) infants born prior to 29 weeks of gestation or 77% in infants born at less than 32 weeks of gestation with a birth weight below 1 kg [3, 6, 7]. Despite significant improvements in perinatal care, that is, surfactant treatment, antenatal corticosteroid treatment, and improvement of invasive and noninvasive ventilation strategies, the incidence of long-term sequelae associated with the disease remained unchanged or increased among the most immature infants [8]. This is presumably due to a significant reduction of mortality rates and an increase in the overall number of successfully treated preterm infants.

Clinically, the infants suffer from impaired respiratory gas exchange, that is, hypoxemia with need for supplemental O_2 and alveolar hypoventilation with resultant hypercapnia, leading to a mismatch of ventilation and perfusion [9].

A variety of risk factors contribute to the development of the disease and result in the characteristic histopathologic changes of impaired alveolarization accompanied by a decrease in small vessel development [1]. These changes are accompanied by characteristic inflammatory changes and extensive extracellular matrix (ECM) remodeling together with increased smooth muscle in small pulmonary arteries and airways [10]. When compared to cases treated in the presurfactant era, the amount of interstitial fibrosis is

Molecular Aspects of Aging: Understanding Lung Aging, First Edition. Edited by Mauricio Rojas, Silke Meiners and Claude Jourdan Le Saux.

© 2014 John Wiley & Sons, Inc. Published 2014 by John Wiley & Sons, Inc.

substantially less and tends to be more diffuse [10]. The change of the histopathologic features characterizing BPD over time reflects the need to consider the initial time of diagnosis when discussing long-term effects.

Increased evidence now suggests that these early pathologic changes of the developing lung contribute to premature aging of the lung and chronic lung disease. The following sections will highlight potential indicators for early aging following pulmonary injury during organ development including functional patterns, pathway signatures, and prenatal or genetic risk factors.

10.2 Altered pulmonary function in infants with BPD

With increasing survival of infants with BPD, attempts to minimize long-term pulmonary impairment and the neurologic complications associated with BPD have become the main focus of perinatal care [11, 12].

Early pulmonary dysfunction is characterized by diminished lung compliance, tachypnea, and increased minute ventilation and work of breathing with and without oxygen dependency. This clinical picture can furthermore be accompanied by an increase in lung microvascular filtration pressure that may lead to interstitial pulmonary edema. The increased lung vascular resistance, typically associated with impaired responsiveness to inhaled nitric oxide and other vasodilators, can progress to reversible or sustained pulmonary hypertension and right heart failure [13, 14]. Early measurement of lung function provides prognostic information, underlined by the finding that infants with more severe lung disease after birth were more likely to develop moderate/severe BPD at term [15].

When the infant reaches term gestation and diagnosis is made, pulmonary function tests demonstrate increased respiratory tract resistance and hyperreactive airways [16] that can manifest as episodic bronchoconstriction and cyanosis.

Affected infants may remain oxygen dependent for months or years, although only few remain oxygen dependent beyond 2 years of age [17, 18] Oxygen dependency indicates the most severe lung disease, as these infants require hospital readmission twice as often compared to infants who are not oxygen dependent. However, even after outgrowing oxygen dependence, patients with moderate or severe BPD still have more outpatient clinic visits, episodes of wheezing, and need for inhaled therapies [15]

Regarding respiratory symptoms, BPD was found to be a significant risk factor for wheeze and medication requirement (odds ratio 2.7 and 2.4, respectively), with about 20%–30% of infants with BPD having those symptoms at 6 and 12 months of age [19, 20]. Respiratory symptoms remain common at preschool and school age [17, 21], with the most severely affected children remaining symptomatic into adulthood [22].

Overall, infants suffering from BPD have a high readmission rate, with up to 70% requiring a hospital stay and about 30% requiring three readmissions in the first 2 years of life [23]. Lower respiratory tract infections resulting from respiratory syncytial virus remain the major cause for readmission among preterm infants regardless of BPD status [24]. Hospitalization rate then declines after the second year of life [25].

Long-term pulmonary function in BPD patients has been characterized by different studies. Important data were generated by the EPICure study [26], which showed significantly lower peak oxygen consumption, forced expiratory volume at one s (FEV$_1$), and gas transfer for infants born extremely premature at school age, compared to age-matched controls. Furthermore, extremely premature infants achieved significantly lower peak work-load and exhibited higher respiratory rates in combination with lower tidal volumes during peak exercise, whereas their residual capacity was increased at school age. These changes may reflect the effect of hyperinflation due to airway obstruction and/or altered pulmonary chemoreceptor function and suggest the presence of persistent airflow limitations and reductions in alveolar surface area.

Lower lung volumes and decreased gas mixing efficiency during infancy in BPD patients have been confirmed by various studies, reflecting abnormalities in lung growth [27, 28]. Up to 80% of preterm infants, particularly those who presented with wheezing, demonstrated airway obstruction in early childhood and adolescence, the majority of whom were symptomatic [29, 30, 31]. There is evidence that BPD survivors reach a reduced maximal airway function (judging from reductions in FEV1 and FEV1/forced vital capacity) as young adults and some even show a trend toward an early and steeper decline in lung function with age, raising concern that BPD may be a precursor of a COPD-like phenotype later in life [32]. Nonetheless, the incidence of COPD has yet to be defined in this patient cohort [33].

Baraldi et al. extrapolated the possible decline in lung function in patients with BPD with and without additional injury, that is, smoke, but up to now, it remains unknown if the rate of decline will parallel that among healthy individuals or is accelerated with advancing age [34]. The indicated BPD patients suffered from variable airflow limita-tion starting in the first years of life together with little evidence of **catch-up** growth in lung function and FEV1 values not reaching the normal maximal value in early adulthood.

Taken together, the functional data available suggest that early lung injury in this patient cohort leads to abnormalities in lung function (and immunity) in infancy and early childhood. Although respiratory symptoms seem to play a secondary role thereaf-ter in one fraction of the patient cohort, significant pulmonary problems persist in the most severely affected infants. The predisposition for early lung function decline in extremely premature infants is suggested by data obtained in early and later adult-hood. Here, measurements of lung function are one important indicator and potential predictor of later lung disease and early lung function decline resulting from early impaired lung development. If performed early on and throughout life, they have the potential to not only reveal initial risk factors but also to elucidate the impact of second-ary injuries, for example, first- and second-hand smoke or viral infections. However, abnormalities of lung function must be interpreted in the context of the era in which BPD was diagnosed, that is, taking into account both the definition of BPD used to define the study population and the standards of perinatal care including exogenous surfactant treatment [1, 35, 36].

10.3 Response to injury

Large clinical trials have identified numerous risk factors for the development of BPD [37, 38, 39, 40, 41, 42]. In addition, experimental studies in animal models were instru-mental in elucidating some of the underlying mechanisms by which these risk factors resulted in profound and durable structural changes in the developing lung. The resulting pathophysiological changes to the lung include alteration of central growth factor signaling pathways in association with profound structural defects. Postnatally, the lung is exposed to injury in the canalicular and saccular stage of organ development, long before defined alveolar structures are present. Prenatally, effects may even occur in more immature stages of pulmonary development.

Most importantly, despite significant improvements in perinatal care, preterm infants often require prolonged assisted ventilation to treat acute respiratory failure caused by primary surfactant deficiency, that is, respiratory distress syndrome. Subsequently, mechanical ventilation and hyperoxia impact the development of a structural and functional immature lung and act beyond the individual genetic background. Here, shear stress and oxygen toxicity exhibit both a direct impact on the pulmonary scaffold and secondary effects resulting from the inflammatory response provoked by these stimuli as also summarized in Figure 10.1 [43, 44, 45, 46]. Although tightly related to each other, the sections 10.3.1–10.3.4 are designed to summarize the findings according to their main biological effects.

10.3.1 Oxidative stress response

Relative deficiencies of antioxidants and inhibitors of proteolytic enzymes render the very immature lung especially vulnerable to the effects of toxic oxygen metabolites and proteases released by the ECM, resident lung cells, or activated neutrophils and macrophages [47, 48, 49, 50]. Different markers have been investigated to indicate enhanced oxidative stress in the preterm infant. Elevated urinary malondialdehyde concentrations in the first week of life, generated by peroxidation of lipid membranes after oxidant-mediated injury, were correlated with the risk for oxygen radical diseases including BPD [51]. Pulmonary antioxidant concentrations have been measured in the lavage of preterm infants; however, they alone were found to be poor predictors of later BPD [52]. Interestingly, recent studies have shown that adolescent BPD patients have evidence of heightened oxidative stress in the airway, suggesting that long-term respiratory abnormalities after preterm birth may be associated with a sustained alteration of the oxidative stress response [53].

Maturational differences exist between neonatal and adult lung cells in terms of the response to apoptotic or proliferative stimuli. While chronic oxygen exposure (60% for 14 days) enhances lung vascular and airway smooth muscle contraction and reduces nitric oxide relaxation in the neonatal rat lung, the opposite occurs in the adult animal [54]. Long-term effects of hyperoxia in the first week of life (100% for 4 days) have been studied and shown to increase mortality by inducing pulmonary vascular disease in mice [55]. These changes are associated with altered lung function (increased compliance) and right ventricular hypertrophy, indicating significant pulmonary hypertension at 67 weeks of age. At this late stage, bone morphogenetic protein signaling is altered and may contribute to the phenotype observed in the adult lung. Regarding other potential mechanisms for the increased susceptibility of the newborn lung, Balasubramaniam et al. showed that hyperoxia reduces bone marrow, circulating, and lung endothelial progenitor cells in the developing lung but not in adult mice [56]. In addition, hyperoxia has been shown to impair alveolar formation and induce senescence through decreased histone deacetylase activity and upregulation of p21 in neonatal mouse lung [57]. Hyperoxia may also alter lung immune function, as studies have shown that different doses of neonatal oxygen in mice affect both,

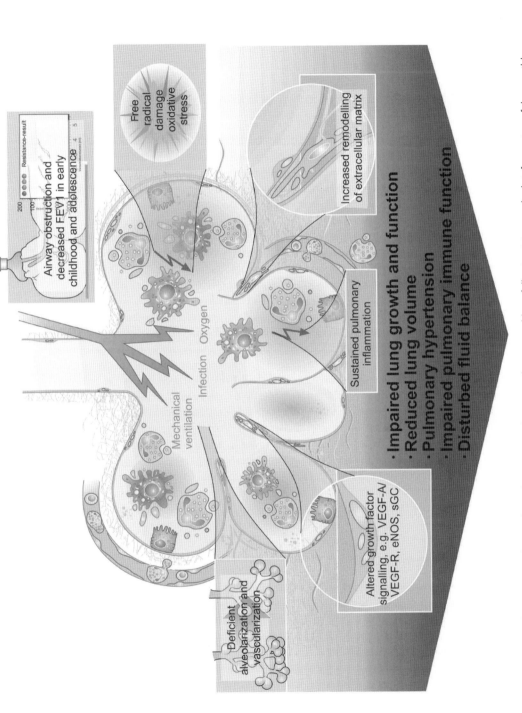

Figure 10.1 Summary of mechanisms that mediate characteristic changes to the neonatal lung following injury and a brief overview of the possible consequences. To see a color version of this figure, see Plate 10.1.

lung development and host response to influenza A virus [58]. A possible mechanistic link mediating neonatal programming of hypoxic sensitivity and the ensuing disruption of cardiorespiratory homeostasis is implicated by a hitherto uncharacterized role for DNA methylation [59].

10.3.2 Extracellular matrix remodeling

Direct effects of shear stress and oxygen toxicity as well as the resulting characteristic inflammatory response lead to the degradation of lung elastin in infants who later acquire BPD. This manifests in increased urinary excretion of desmosin, a breakdown product of the mature elastic fiber, preceded and paralleled by an associated increase in protease activity [60, 61, 62]. Similarly, the increase in desmosin was found to predict disease severity and outcome in adult patients with acute respiratory distress syndrome (ARDS) [63], highlighting the importance of ECM degradation irrespective of the pulmonary developmental stage. Maintaining a delicate balance of protease activity is necessary for normal lung development, as evidenced by studies showing that complete matrix metalloproteinase deficiency worsens lung injury in an inflammatory model of BPD, potentially due to impaired defense mechanisms or decreased migratory potential of various cell types, important for normal lung development and repair [64].

Increased remodeling of the ECM results in both a greater abundance and abnormal distribution of elastin and deformation of collagen scaffolding in the pulmonary parenchyma and circulation, shown both in infants with BPD and in authentic animal models [65, 66, 67]. Studies suggest the underlying mechanisms may include factors such as volutrauma, hyperoxia, and inflammation that could disrupt the parathyroid hormone-related protein receptor pathway from endoderm to mesoderm, causing maladaptive changes that may lead to myofibroblast proliferation [68]. Changes to the structural integrity of the ECM potentially affect its function as a scaffold for the formation of new alveoli and capillaries. The ECM has been shown to hold a memory function that is involved in defining the fate of the cells that populate the lung [69, 70]. Thus, its sustained and irreversible reorganization may result in long-term changes to the lung, affecting the development of new alveoli, repair after injury, or even the binding of inflammatory cells invading the lung during infections.

10.3.3 Inflammation

Pulmonary inflammation, induced both by noninfectious agents, such as positive-pressure ventilation, oxygen therapy, primary surfactant deficiency, or patent ductus arteriosus, and pre- or postnatal infections plays an important role in **priming** the immature lung to susceptibility for the development of BPD [40, 44, 71, 72]. The range of inflammation and infections includes the prenatal impact of cytokines (i.e., fetal inflammatory response syndrome), chorioamnionitis, as well as congenital and nosocomial infections. The characteristic influx of neutrophils into the lung is accompanied by increased numbers of macrophages in the course of the disease [73, 74, 75]. Animal studies indicate that lung injury leading to ECM remodeling or early alveolar epithelial dysfunction promotes lung inflammation [76, 77], giving a potential link for early origin of later (impaired) pulmonary immune function.

10.3.4 Morphogenetic response

The release of cytokines and altered growth factor signaling (e.g., transforming growth factor (TGF)-β) lead to the activation of different transcription factors and result in a characteristic increase in apoptosis affecting all different cell types [78]. Other important growth factors include the platelet-derived growth factor-α and the fibroblast growth factor family (for a review, see [79]). Early interference with different transcription factors disrupts normal lung morphogenesis in fetal life [80], subsequently driving the onset of severe chronic bronchial inflammation, resulting in pulmonary emphysema in adult mice [81]. Taken together, these altered signaling pathways not only affect normal lung development but also likely impact the pulmonary response to injury and repair mechanisms long term. Here, the developing lung differs from the well-known response of adult pulmonary tissue. Thus, suppression of the nuclear factor kappa B (NF-kB) has been shown to not prevent but worsen pathophysiological changes that potentially lead to BPD development, for example, alveolarization and vascularization [80, 82, 83].

The process of deficient alveolarization is characteristically associated with the presence of dysmorphic capillaries and an altered pattern of angiogenic growth factors, thereby contributing to subsequent development of pulmonary hypertension and impaired lung lymphatic drainage [13, 14]. Pulmonary expression of vascular endothelial growth factor (VEGF) and VEGF receptors [84, 85, 86] is reduced, accompanied by diminished endothelial nitric oxide synthase (eNOS) and soluble guanylate cyclase (sGC) in lung blood vessels and airways [87, 88]. These changes reflect the expression pattern observed in aged mice [89] and very likely contribute to reduced plasticity of lung capillaries [90].

10.4 Prenatal and genetic predisposition

Prenatal risk factors influence the capacity of the developing lung to respond to injuries such as stretch, oxygen, infection, and inflammation. Intrauterine growth retardation increases the risk of BPD three- to fourfold [75, 91, 92, 93, 94], most likely through impaired alveolar and vascular growth associated with altered growth factor signaling [95]. Poor nutritional support, vitamin deficiency, as well as insufficient adrenal and thyroid hormone release in the very premature infant further increase the risk after birth [96, 97, 98].

Therapeutically, the prenatal administration of betamethasone, widely used to enhance lung maturation and to prevent respiratory distress, has been shown to reduce BPD rates [99, 100]. Nonetheless, antenatal betamethasone has been shown to be associated with an increase in indicators of lipid membrane peroxidation [51]. Postnatal dexamethasone treatment in the neonate had adverse effects that also manifested in elderly rats, that is, systolic dysfunction and reduced life expectancy [101, 102]. These findings suggest a careful investigation of long-term treatment effects.

The broad use of antibiotic treatment in the mother at risk for premature birth alters the bacterial flora of the child [103], affecting immune function long term as shown in neonatal – but not adult – germ-free mice [104].

Regarding the impact of the genetic background, a study investigating twin preterm infants found genetic factors to account for 53% of the variance in liability for BPD [105]. Several potential candidate genes have been shown in association with BPD, where genetic variants predisposing to BPD are usually polymorphisms which are not causative, but can increase susceptibility to the disease. Genetic abnormalities include variations affecting the surfactant system or variables of the innate immune response [106, 107]. Equally important, the identification of genetic associations often serves to identify novel pathways implicated in the disease, that is, the association of superoxide dismutase mutations with the development of BPD [108].

In addition, male preterm infants are at a higher risk for the development of long-term impairment, including BPD [109], and premature changes to hormonal regulation have been discussed as an underlying cause [110]. The effect of gender seems to be different with respect to the adult population, as female adult BPD patients are more severely affected with respect to developing long-term pulmonary impairment [111].

10.5 Conclusion

To conclude, the unique response to injury observed in the immature lung with respect to oxidative stress, ECM composition, growth factor signaling, and the initiation of a sustained inflammatory response results in a characteristic histopathologic picture that is known to translate into pulmonary long-term consequences as summarized in Figure 10.1. We now understand that the observed early organ injury provokes a characteristic response to later challenges and physiologic aging processes and thereby reflects a pulmonary **memory effect** rather than the effects being erased over time. This and the possible effects of an altered pulmonary system on other developing organs, that is, the brain, have to be taken into account when treatment strategies are designed and lifestyle issues are advocated to this patient population.

References

1. Jobe, A. H. and E. Bancalari. 2001. Bronchopulmonary dysplasia. *Am J Respir Crit Care Med* 163(7): 1723–9.
2. Gortner, L., B. Misselwitz, D. Milligan, et al. 2011. Rates of bronchopulmonary dysplasia in very preterm neonates in Europe: results from the MOSAIC cohort. *Neonatology* 99(2):112–7.
3. Stoll, B. J., N. I. Hansen, E. F. Bell, et al. 2010. Neonatal outcomes of extremely preterm infants from the NICHD Neonatal Research Network. *Pediatrics* 126(3):443–56.
4. Vachon, E., Y. Bourbonnais, C. D. Bingle, S. J. Rowe, M. F. Janelle, and G. M. Tremblay. 2002. Anti-inflammatory effect of pre-elafin in lipopolysaccharide-induced acute lung inflammation. *Biol Chem* 383(7–8):1249–56.
5. Van Marter, L. J., M. Pagano, E. N. Allred, A. Leviton, and K. C. Kuban. 1992. Rate of bronchopulmonary dysplasia as a function of neonatal intensive care practices. *J Pediatr* 120(6):938–46.
6. Johnson, A. H., J. L. Peacock, A. Greenough, et al. 2002. High-frequency oscillatory ventilation for the prevention of chronic lung disease of prematurity. *N Engl J Med* 347(9):633–42.
7. Ehrenkranz, R. A., M. C. Walsh, B. R. Vohr, et al. 2005. Validation of the National Institutes of Health consensus definition of bronchopulmonary dysplasia. *Pediatrics* 116(6):1353–60.
8. Doyle, L. W. 2006. Evaluation of neonatal intensive care for extremely-low-birth-weight infants. *Semin Fetal Neonatal Med* 11(2):139–45.
9. Lopez, E., J. Mathlouthi, S. Lescure, B. Krauss, P. H. Jarreau, and G. Moriette. Capnography in spontaneously breathing preterm infants with bronchopulmonary dysplasia. *Pediatr Pulmonol* 46(9):896–902.
10. Husain, A. N., N. H. Siddiqui, and J. T. Stocker. 1998. Pathology of arrested acinar development in postsurfactant bronchopulmonary dysplasia. *Hum Pathol* 29(7):710–7.
11. Jobe, A. H. and M. Ikegami. 2001. Prevention of bronchopulmonary dysplasia. *Curr Opin Pediatr* 13(2):124–9.

12. Newman, J. B., A. G. Debastos, D. Batton, and S. Raz. 2011. Neonatal respiratory dysfunction and neuropsychological performance at the preschool age: a study of very preterm infants with bronchopulmonary dysplasia. *Neuropsychology* 25(5):666–78.

13. Kinsella, J. P., A. Greenough, and S. H. Abman. 2006. Bronchopulmonary dysplasia. *Lancet* 367(9520):1421–31.

14. Steinhorn, R. H. Neonatal pulmonary hypertension. *Pediatr Crit Care Med* 11(2 Suppl):S79–84.

15. May, C., C. Kennedy, A. D. Milner, G. F. Rafferty, J. L. Peacock, and A. Greenough. 2011. Lung function abnormalities in infants developing bronchopulmonary dysplasia. *Arch Dis Child* 96(11):1014–9.

16. Hilgendorff, A., I. Reiss, L. Gortner, D. Schuler, K. Weber, and H. Lindemann. 2008. Impact of airway obstruction on lung function in very preterm infants at term. *Pediatr Crit Care Med* 9(6):629–35.

17. Greenough, A., J. Alexander, S. Burgess, et al. 2006. Preschool healthcare utilisation related to home oxygen status. *Arch Dis Child Fetal Neonatal Ed* 91(5):F337–41.

18. Greenough, A., J. Alexander, S. Burgess, P. A. et al. 2002. Home oxygen status and rehospitalisation and primary care requirements of infants with chronic lung disease. *Arch Dis Child* 86(1):40–3.

19. Greenough, A., E. Limb, L. Marston, N. Marlow, S. Calvert, and J. Peacock. 2005. Risk factors for respiratory morbidity in infancy after very premature birth. *Arch Dis Child Fetal Neonatal Ed* 90(4):F320–3.

20. Baraldi, E., S. Carraro, and M. Filippone. 2009. Bronchopulmonary dysplasia: definitions and long-term respiratory outcome. *Early Hum Dev* 85(10 Suppl):S1–3.

21. Gross, S. J., D. M. Iannuzzi, D. A. Kveselis, and R. D. Anbar. 1998. Effect of preterm birth on pulmonary function at school age: a prospective controlled study. *J Pediatr* 133(2):188–92.

22. Northway, W. H., Jr., R. B. Moss, K. B. Carlisle, et al. 1990. Late pulmonary sequelae of bronchopulmonary dysplasia. *N Engl J Med* 323(26):1793–9.

23. Greenough, A., S. Cox, J. Alexander, et al. 2001. Health care utilisation of infants with chronic lung disease, related to hospitalisation for RSV infection. *Arch Dis Child* 85(6):463–8.

24. Broughton, S., A. Roberts, G. Fox, et al. 2005. Prospective study of healthcare utilisation and respiratory morbidity due to RSV infection in prematurely born infants. *Thorax* 60(12):1039–44.

25. Doyle, L. W., M. M. Cheung, G. W. Ford, A. Olinsky, N. M. Davis, and C. Callanan. 2001. Birthweight <1501 g and respiratory health at age 14. *Arch Dis Child* 84(1):40–4.

26. Welsh, L., J. Kirkby, S. Lum, et al. 2010. The EPICure study: maximal exercise and physical activity in school children born extremely preterm. *Thorax* 65(2):165–72.

27. Greenough, A., G. Dimitriou, R. Y. Bhat, et al. 2005. Lung volumes in infants who had mild to moderate bronchopulmonary dysplasia. *Eur J Pediatr* 164(9):583–6.

28. Hjalmarson, O. and K. L. Sandberg. 2005. Lung function at term reflects severity of bronchopulmonary dysplasia. *J Pediatr* 146(1):86–90.

29. Broughton, S., M. R. Thomas, L. Marston, et al. 2007. Very prematurely born infants wheezing at follow-up: lung function and risk factors. *Arch Dis Child* 92(9):776–80.

30. Yuksel, B. and A. Greenough. 1991. Relationship of symptoms to lung function abnormalities in preterm infants at follow-up. *Pediatr Pulmonol* 11(3):202–6.

31. Pelkonen, A. S., A. L. Hakulinen, and M. Turpeinen. 1997. Bronchial lability and responsiveness in school children born very preterm. *Am J Respir Crit Care Med* 156(4 Pt 1):1178–84.

32. Doyle, L. W., B. Faber, C. Callanan, N. Freezer, G. W. Ford, and N. M. Davis. 2006. Bronchopulmonary dysplasia in very low birth weight subjects and lung function in late adolescence. *Pediatrics* 118(1):108–13.

33. Filippone, M., S. Carraro, and E. Baraldi. 2010. From BPD to COPD? The hypothesis is intriguing but we lack lung pathology data in humans. *Eur Respir J* 35(6):1419–20; author reply 1420.

34. Baraldi, E. and M. Filippone. 2007. Chronic lung disease after premature birth. *N Engl J Med* 357(19):1946–55.

35. Jobe, A. H. 2011. The new bronchopulmonary dysplasia. *Curr Opin Pediatr* 23(2):167–72.

36. Northway, W. H., Jr., R. C. Rosan, and D. Y. Porter. 1967. Pulmonary disease following respirator therapy of hyaline-membrane disease. Bronchopulmonary dysplasia. *N Engl J Med* 276(7): 357–68.

37. Korhonen, P., O. Tammela, A. M. Koivisto, P. Laippala, and S. Ikonen. 1999. Frequency and risk factors in bronchopulmonary dysplasia in a cohort of very low birth weight infants. *Early Hum Dev* 54(3):245–58.

38. Clyman, R., G. Cassady, J. K. Kirklin, M. Collins, and J. B. Philips, 3rd. 2009. The role of patent ductus arteriosus ligation in bronchopulmonary dysplasia: reexamining a randomized controlled trial. *J Pediatr* 154(6):873–6.

39. Oh, W., B. B. Poindexter, R. Perritt, et al. 2005. Association between fluid intake and weight loss during the first ten days of life and risk of bronchopulmonary dysplasia in extremely low birth weight infants. *J Pediatr* 147(6):786–90.

40. Kramer, B. W. 2008. Antenatal inflammation and lung injury: prenatal origin of neonatal disease. *J Perinatol* 28(Suppl 1):S21–7.

41. Stevens, T. P., E. W. Harrington, M. Blennow, and R. F. Soll. 2007. Early surfactant administration with brief ventilation vs. selective surfactant and continued mechanical ventilation for preterm infants with or at risk for respiratory distress syndrome. *Cochrane Database Syst Rev* (4):CD003063.

42. Mittendorf, R., R. Covert, A. G. Montag, et al. 2005. Special relationships between fetal inflammatory response syndrome and bronchopulmonary dysplasia in neonates. *J Perinat Med* 33(5):428–34.

43. Jobe, A. H. and M. Ikegami. 1998. Mechanisms initiating lung injury in the preterm. *Early Hum Dev* 53(1):81–94.

44. Watterberg, K. L., L. M. Demers, S. M. Scott, and S. Murphy. 1996. Chorioamnionitis and early lung inflammation in infants in whom bronchopulmonary dysplasia develops. *Pediatrics* 97(2):210–5.

45. Hislop, A. A., J. S. Wigglesworth, R. Desai, and V. Aber. 1987. The effects of preterm delivery and mechanical ventilation on human lung growth. *Early Hum Dev* 15(3):147–64.

46. Brew, N., S. B. Hooper, B. J. Allison, M. J. Wallace, and R. Harding. 2011. Injury and repair in the very immature lung following brief mechanical ventilation. *Am J Physiol Lung Cell Mol Physiol* 301(6):L917–26.

47. Rose, M. J., M. R. Stenger, M. S. Joshi, S. E. Welty, J. A. Bauer, and L. D. Nelin. Inhaled nitric oxide decreases leukocyte trafficking in the neonatal mouse lung during exposure to >95% oxygen. *Pediatr Res* 67(3):244–9.

48. Bose, C. L., C. E. Dammann, and M. M. Laughon. 2008. Bronchopulmonary dysplasia and inflammatory biomarkers in the premature neonate. *Arch Dis Child Fetal Neonatal Ed* 93(6): F455–61.

49. Vento, G., C. Tirone, P. Lulli, et al. 2009. Bronchoalveolar lavage fluid peptidomics suggests a possible matrix metalloproteinase-3 role in bronchopulmonary dysplasia. *Intensive Care Med* 35(12):2115–24.

50. Watterberg, K. L., D. F. Carmichael, J. S. Gerdes, S. Werner, C. Backstrom, and S. Murphy. 1994. Secretory leukocyte protease inhibitor and lung inflammation in developing bronchopulmonary dysplasia. *J Pediatr* 125(2):264–9.

51. Weinberger, B., M. Anwar, S. Henien, et al. 2004. Association of lipid peroxidation with antenatal betamethasone and oxygen radial disorders in preterm infants. *Biol Neonate* 85(2):121–7.

52. Collard, K. J., S. Godeck, J. E. Holley, and M. W. Quinn. 2004. Pulmonary antioxidant concentrations and oxidative damage in ventilated premature babies. *Arch Dis Child Fetal Neonatal Ed* 89(5):F412–6.

53. Filippone, M., G. Bonetto, M. Corradi, A. C. Frigo, and E. Baraldi. 2012. Evidence of unexpected oxidative stress in airways of adolescents born very pre-term. *Eur Respir J* 40(5):1253–9.

54. Belik, J., R. P. Jankov, J. Pan, and A. K. Tanswell. 2003. Chronic O_2 exposure enhances vascular and airway smooth muscle contraction in the newborn but not adult rat. *J Appl Physiol* 94(6):2303–12.

55. Yee, M., R. J. White, H. A. Awad, W. A. Bates, S. A. McGrath-Morrow, and M. A. O'Reilly. 2011. Neonatal hyperoxia causes pulmonary vascular disease and shortens life span in aging mice. *Am J Pathol* 178(6):2601–10.

56. Balasubramaniam, V., C. F. Mervis, A. M. Maxey, N. E. Markham, and S. H. Abman. 2007. Hyperoxia reduces bone marrow, circulating, and lung endothelial progenitor cells in the developing lung: implications for the pathogenesis of bronchopulmonary dysplasia. *Am J Physiol Lung Cell Mol Physiol* 292(5):L1073–84.

57. Londhe, V. A., I. K. Sundar, B. Lopez, et al. 2011. Hyperoxia impairs alveolar formation and induces senescence through decreased histone deacetylase activity and up-regulation of p21 in neonatal mouse lung. *Pediatr Res* 69(5 Pt 1):371–7.

58. Buczynski, B. W., M. Yee, B. Paige Lawrence, and M. A. O'Reilly. 2012. Lung development and the host response to influenza A virus are altered by different doses of neonatal oxygen in mice. *Am J Physiol Lung Cell Mol Physiol* 302(10):L1078–87.

59. Nanduri, J., V. Makarenko, V. D. Reddy, et al. 2012. Epigenetic regulation of hypoxic sensing disrupts cardiorespiratory homeostasis. *Proc Natl Acad Sci USA* 109(7):2515–20.

60. Bruce, M. C., M. Schuyler, R. J. Martin, B. C. Starcher, J. F. Tomashefski, Jr., and K. E. Wedig. 1992. Risk factors for the degradation of lung elastic fibers in the ventilated neonate. Implications for impaired lung development in bronchopulmonary dysplasia. *Am Rev Respir Dis* 146(1):204–12.

61. Bruce, M. C., K. E. Wedig, N. Jentoft, et al. 1985. Altered urinary excretion of elastin cross-links in premature infants who develop bronchopulmonary dysplasia. *Am Rev Respir Dis* 131(4):568–72.

62. Merritt, T. A., C. G. Cochrane, K. Holcomb, et al. 1983. Elastase and alpha 1-proteinase inhibitor activity in tracheal aspirates during respiratory distress syndrome. Role of inflammation in the pathogenesis of bronchopulmonary dysplasia. *J Clin Invest* 72(2):656–66.

63. Tenholder, M. F., K. R. Rajagopal, Y. Y. Phillips, et al. 1991. Urinary desmosine excretion as a marker of lung injury in the adult respiratory distress syndrome. *Chest* 100(5):1385–90.

64. Lukkarinen, H., A. Hogmalm, U. Lappalainen, and K. Bry. 2009. Matrix metalloproteinase-9 deficiency worsens lung injury in a model of bronchopulmonary dysplasia. *Am J Respir Cell Mol Biol* 41(1):59–68.

65. Pierce, R. A., K. H. Albertine, B. C. Starcher, J. F. Bohnsack, D. P. Carlton, and R. D. Bland. 1997. Chronic lung injury in preterm lambs: disordered pulmonary elastin deposition. *Am J Physiol* 272(3 Pt 1):L452–60.

66. Thibeault, D. W., S. M. Mabry, Ekekezie, II, and W. E. Truog. 2000. Lung elastic tissue maturation and perturbations during the evolution of chronic lung disease. *Pediatrics* 106(6):1452–9.

67. Thibeault, D. W., S. M. Mabry, Ekekezie, II, X. Zhang, and W. E. Truog. 2003. Collagen scaffolding during development and its deformation with chronic lung disease. *Pediatrics* 111(4 Pt 1):766–76.

68. Torday, J. S. and V. K. Rehan. 2003. Mechanotransduction determines the structure and function of lung and bone: a theoretical model for the pathophysiology of chronic disease. *Cell Biochem Biophys* 37(3):235–46.

69. Bonvillain, R. W., S. Danchuk, D. E. Sullivan, et al. 2012. A nonhuman primate model of lung regeneration: detergent-mediated decellularization and initial in vitro recellularization with mesenchymal stem cells. *Tissue Eng Part A* 18(23–24):2437–52.

70. Jensen, T., B. Roszell, F. Zang, et al. 2012. A rapid lung de-cellularization protocol supports embryonic stem cell differentiation in vitro and following implantation. *Tissue Eng Part C Methods* 18(8):632–46.

71. Stoll, B. J., T. Gordon, S. B. Korones, et al. 1996. Early-onset sepsis in very low birth weight neonates: a report from the National Institute of Child Health and Human Development Neonatal Research Network. *J Pediatr* 129(1):72–80.

72. Yoon, B. H., R. Romero, J. K. Jun, et al. 1997. Amniotic fluid cytokines (interleukin-6, tumor necrosis factor-alpha, interleukin-1 beta, and interleukin-8) and the risk for the development of bronchopulmonary dysplasia. *Am J Obstet Gynecol* 177(4):825–30.

73. Speer, C. P. 2006. Inflammation and bronchopulmonary dysplasia: a continuing story. *Semin Fetal Neonatal Med* 12:12.

74. Todd, D. A., M. Earl, J. Lloyd, M. Greenberg, and E. John. 1998. Cytological changes in endotracheal aspirates associated with chronic lung disease. *Early Hum Dev* 51(1):13–22.

75. Walsh, M. C., Q. Yao, J. D. Horbar, J. H. Carpenter, S. K. Lee, and A. Ohlsson. 2006. Changes in the use of postnatal steroids for bronchopulmonary dysplasia in 3 large neonatal networks. *Pediatrics* 118(5):e1328–35.

76. Atochina-Vasserman, E. N., S. R. Bates, P. Zhang, et al. 2011. Early alveolar epithelial dysfunction promotes lung inflammation in a mouse model of Hermansky-Pudlak syndrome. *Am J Respir Crit Care Med* 184(4):449–58.

77. Hilgendorff, A., K. Parai, R. Ertsey, et al. 2011. Inhibiting lung elastase activity enables lung growth in mechanically ventilated newborn mice. *Am J Respir Crit Care Med* 184(5):537–46.

78. Kunzmann, S., C. P. Speer, A. H. Jobe, and B. W. Kramer. 2007. Antenatal inflammation induced TGF-beta1 but suppressed CTGF in preterm lungs. *Am J Physiol Lung Cell Mol Physiol* 292(1):L223–31.

79. Jankov, R. P. and A. K. Tanswell. 2004. Growth factors, postnatal lung growth and bronchopulmonary dysplasia. *Pediatr Respir Rev* 5(Suppl A):S265–275.

80. Iosef, C., T. P. Alastalo, Y. Hou, et al. 2012. Inhibiting NF-kappaB in the developing lung disrupts angiogenesis and alveolarization. *Am J Physiol Lung Cell Mol Physiol* 302(10):L1023–36.

81. Didon, L., A. B. Roos, G. P. Elmberger, F. J. Gonzalez, and M. Nord. 2010. Lung-specific inactivation of CCAAT/enhancer binding protein alpha causes a pathological pattern characteristic of COPD. *Eur Respir J* 35(1):186–97.

82. Alvira, C. M., A. Abate, G. Yang, P. A. Dennery, and M. Rabinovitch. 2007. Nuclear factor-kappaB activation in neonatal mouse lung protects against lipopolysaccharide-induced inflammation. *Am J Respir Crit Care Med* 175(8):805–15.

83. Yang, G., A. Abate, A. G. George, Y. H. Weng, and P. A. Dennery. 2004. Maturational differences in lung NF-kappaB activation and their role in tolerance to hyperoxia. *J Clin Invest* 114(5):669–78.

84. Thebaud, B. 2007. Angiogenesis in lung development, injury and repair: implications for chronic lung disease of prematurity. *Neonatology* 91(4):291–7.

85. De Paepe, M. E., D. Greco, and Q. Mao. 2010. Angiogenesis-related gene expression profiling in ventilated preterm human lungs. *Exp Lung Res* 36(7):399–410.

86. De Paepe, M. E., Q. Mao, J. Powell, et al. 2006. Growth of pulmonary microvasculature in ventilated preterm infants. *Am J Respir Crit Care Med* 173(2):204–11.

87. Bland, R. D., C. Y. Ling, K. H. Albertine, et al. 2003. Pulmonary vascular dysfunction in preterm lambs with chronic lung disease. *Am J Physiol Lung Cell Mol Physiol* 285(1):L76–85.

88. Vyas-Read, S., P. W. Shaul, I. S. Yuhanna, and B. C. Willis. 2007. Nitric oxide attenuates epithelial-mesenchymal transition in alveolar epithelial cells. *Am J Physiol Lung Cell Mol Physiol* 293(1):L212–21.

89. Ito, Y., T. Betsuyaku, K. Nagai, Y. Nasuhara, and M. Nishimura. 2005. Expression of pulmonary VEGF family declines with age and is further down-regulated in lipopolysaccharide (LPS)-induced lung injury. *Exp Gerontol* 40(4):315–23.

90. Kamba, T., B. Y. Tam, H. Hashizume, et al. 2006. VEGF-dependent plasticity of fenestrated capillaries in the normal adult microvasculature. *Am J Physiol Heart Circ Physiol* 290(2):H560–76.

91. Regev, R. H., A. Lusky, T. Dolfin, I. Litmanovitz, S. Arnon, and B. Reichman. 2003. Excess mortality and morbidity among small-for-gestational-age premature infants: a population-based study. *J Pediatr* 143(2):186–91.

92. Reiss, I., E. Landmann, M. Heckmann, B. Misselwitz, and L. Gortner. 2003. Increased risk of bronchopulmonary dysplasia and increased mortality in very preterm infants being small for gestational age. *Arch Gynecol Obstet* 269(1):40–4.

93. Rieger-Fackeldey, E., A. Schulze, F. Pohlandt, R. Schwarze, J. Dinger, and W. Lindner. 2005. Short-term outcome in infants with a birthweight less than 501 grams. *Acta Paediatr* 94(2):211–6.

94. Bose, C., L. J. Van Marter, M. Laughon, et al. 2009. Fetal growth restriction and chronic lung disease among infants born before the 28th week of gestation. *Pediatrics* 124(3):e450–8.

95. Rozance, P. J., G. J. Seedorf, A. Brown, et al. 2011. Intrauterine growth restriction decreases pulmonary alveolar and vessel growth and causes pulmonary artery endothelial cell dysfunction in vitro in fetal sheep. *Am J Physiol Lung Cell Mol Physiol* 301(6):L860–71.

96. Biniwale, M. A. and R. A. Ehrenkranz. 2006. The role of nutrition in the prevention and management of bronchopulmonary dysplasia. *Semin Perinatol* 30(4):200–8.

97. Shenai, J. P., F. Chytil, and M. T. Stahlman. 1985. Vitamin A status of neonates with bronchopulmonary dysplasia. *Pediatr Res* 19(2):185–8.

98. Watterberg, K. L. and S. M. Scott. 1995. Evidence of early adrenal insufficiency in babies who develop bronchopulmonary dysplasia. *Pediatrics* 95(1):120–5.

99. Hennessy, E. M., M. A. Bracewell, N. Wood, et al. 2008. Respiratory health in pre-school and school age children following extremely preterm birth. *Arch Dis Child* 93(12):1037–43.

100. Merritt, T. A., D. D. Deming, and B. R. Boynton. 2009. The "new" bronchopulmonary dysplasia: challenges and commentary. *Semin Fetal Neonatal Med* 14(6):345–57.

101. Bal, M. P., W. B. de Vries, M. F. van Oosterhout, et al. 2008. Long-term cardiovascular effects of neonatal dexamethasone treatment: hemodynamic follow-up by left ventricular pressure-volume loops in rats. *J Appl Physiol* 104(2):446–50.

102. Kamphuis, P. J., W. B. de Vries, J. M. Bakker, et al. 2007. Reduced life expectancy in rats after neonatal dexamethasone treatment. *Pediatr Res* 61(1):72–6.

103. Stoll, B. J., N. Hansen, A. A. Fanaroff, et al. 2002. Changes in pathogens causing early-onset sepsis in very-low-birth-weight infants. *N Engl J Med* 347(4):240–7.

104. Olszak, T., D. An, S. Zeissig, et al. 2012. Microbial exposure during early life has persistent effects on natural killer T cell function. *Science* 336(6080):489–93.

105. Bhandari, V., M. J. Bizzarro, A. Shetty, et al. 2006. Familial and genetic susceptibility to major neonatal morbidities in preterm twins. *Pediatrics* 117(6):1901–6.

106. Hallman, M. and R. Haataja. 2003. Genetic influences and neonatal lung disease. *Semin Neonatol* 8(1):19–27.

107. Hilgendorff, A., K. Heidinger, A. Pfeiffer, et al. 2007. Association of polymorphisms in the mannose-binding lectin gene and pulmonary morbidity in preterm infants. *Genes Immun* 8(8):671–7.
108. Poggi, C., B. Giusti, A. Vestri, E. Pasquini, R. Abbate, and C. Dani. 2012. Genetic polymorphisms of antioxidant enzymes in preterm infants. *J Matern Fetal Neonatal Med* 25(Suppl 4):131–4.
109. Binet, M. E., E. Bujold, F. Lefebvre, Y. Tremblay, and B. Piedboeuf. 2012. Role of gender in morbidity and mortality of extremely premature neonates. *Am J Perinatol* 29(3):159–66.
110. Trotter, A., L. Maier, M. Kron, and F. Pohlandt. 2007. Effect of oestradiol and progesterone replacement on bronchopulmonary dysplasia in extremely preterm infants. *Arch Dis Child Fetal Neonatal Ed* 92(2):F94–8.
111. Vrijlandt, E. J., J. Gerritsen, H. M. Boezen, and E. J. Duiverman. 2005. Gender differences in respiratory symptoms in 19-year-old adults born preterm. *Respir Res* 6:117.

11 Remodeling of the Extracellular Matrix in the Aging Lung

Jesse Roman

Department of Medicine, Division of Pulmonary, Critical Care and Sleep Disorders,
Department of Pharmacology & Toxicology, Robley Rex Veterans
Affairs Medical Center and University of Louisville, Kentucky, USA

11.1 Introduction

With the dreaded appearance of folds and wrinkles, skin changes represent one of the most dramatic manifestations of aging. These changes point to alterations in connective tissue that lead to the characteristic appearance of the aging skin [1]. However, alterations in connective tissue during aging are not limited to the skin. It is well known that the aged or senescent lung is characterized by what has been termed **emphysema of aging**, which is manifested by enlargement of alveolar spaces and loss of alveolar attachments supporting peripheral airways [2, 3]. Although these changes were described years ago, studies investigating these alterations and the mechanisms that drive them are scarce. More importantly, the implications of these changes to lung function remain unclear. Here, we discuss the literature available regarding connective tissue changes in senescent lungs and propose that these changes render the aging host susceptible to infection and injury.

11.2 The aging lung

Aging is associated with increased incidence of respiratory disorders, and elderly patients represent a disproportionate number of afflicted individuals with pneumonia, acute lung injury, and lung fibrosis, among other lung disorders [3, 4, 5]. This association has also been noted in experimental models of lung disease. For example, senescent rodent lungs are more susceptible to lung injury in the setting of mechanical ventilation, ozone exposure, and pulmonary infection [6, 7, 8, 9, 10, 11, 12]. Furthermore, intratracheal instillation of lipopolysaccharide results in the exaggerated expression of proinflammatory cytokines in senescent animals when compared to young controls [8, 13]. Senescent lungs are also more susceptible to bleomycin-induced lung injury [14]. Together, these studies point to the enhanced susceptibility of the senescent lung to injury, but little is known about the factors responsible for this susceptibility.

Molecular Aspects of Aging: Understanding Lung Aging, First Edition. Edited by Mauricio Rojas, Silke Meiners and Claude Jourdan Le Saux.
© 2014 John Wiley & Sons, Inc. Published 2014 by John Wiley & Sons, Inc.

Several mechanisms have been proposed to explain the aforementioned observations including increased oxidant stress and free radical damage, a decline in immune responses, and alterations in stem cell/progenitor cell differentiation potential [6, 15, 16]. Mitochondrial dysfunction has also been implicated in aging since the coordination between nuclear and mitochondrial communication during aging appears to be affected [17]. In addition, a role for dysregulation of the mammalian target of rapamycin (mTOR) pathway has been suggested [18, 19]. Here, we discuss changes in the senescent lung that point to the remodeling of extracellular matrices as a critical determinant of lung function and disease susceptibility.

11.3 Activation of tissue remodeling in the senescent lung

Physiological lung aging is associated with the enlargement of alveolar spaces, the reduced surface area for gas exchange, and the loss of alveolar attachments supporting peripheral airways. These changes are mild but can lead to measurable functional changes characterized by reduced static elastic recoil and decreased compliance of the chest wall and the strength of the respiratory muscles [2, 3]. In a cross-sectional study conducted in 600 healthy nonsmokers aged 19 to 92, investigators found significant decreases in several pulmonary physiological parameters including vital capacity, forced vital capacity, and diffusing capacity [20].

Although the factors that lead to these physiological changes are unclear, they are likely driven, at least in part, by alterations in lung extracellular matrix deposition and composition. In the late 1980s, using immunohistochemistry, investigators found decreased elastic fibers along the alveolar walls, while type III collagen increased in elderly subjects compared to controls [21]. These findings, however, have not been consistent in all studies. For example, when examining the tissues of lungs harvested postmortem after sudden death from subjects aged 15 to 83, others found decreased collagen content but few changes in elastin content [22]. Similar inconsistencies have been found in animal studies. In 20-month-old C57BL/6 mice, researchers found increased lung volume and respiratory system compliance when compared to 2-month-old mice. These changes were associated with decreased elastic fiber content, while collagen content increased [23]. In contrast, other investigators found no changes in elastin content in 24-month-old Balb/c mice compared to 3-month-old animals. Collagen content increased, but there were no significant changes in hydroxyproline content per dry lung weight or the proportion of type III to type I collagen. The authors concluded that in terms of the extracellular matrix, the lungs of aged mice are not very different from the lungs of young mice and that the changes observed were probably the simple consequence of growth [24]. One wonders if the paradoxic observations encountered in these studies are related to the different methods of analysis used and/or genetic differences between strains of mice.

More recently, when examining 2-month-old versus 24-month-old C57BL/6 murine lungs, Sueblinvong et al. found an increase in the expression of mRNA and protein for transforming growth factor-β (TGF-β) [14], a profibrotic growth factor considered a master switch for lung injury, disrepair, and fibrotic pathways in many organs [25, 26, 27]. Among the many actions of TGF-β is the induction of genes coding for molecules involved in tissue remodeling ranging from matrix glycoproteins and fibrillar collagens to proteoglycans and matrix metalloproteinases (MMPs) [27]. The investigators also found increased expression of mRNA coding for Smad3, a transcription factor that mediates many, but not all, of the cellular effects of TGF-β [28]. These findings suggest that growth factor-mediated signaling pathways responsible for tissue remodeling are activated in senescent lungs.

To evaluate for downstream consequences of excess TGF-β, Sueblinvong and colleagues examined for alterations in extracellular matrices; these studies documented increased expression of fibronectin and, specifically, the fibronectin EDA (for extra domain A) splice variant. Fibronectin is a matrix glycoprotein highly expressed in embryonic tissues and considered important for lung development [29]. After birth, fibronectin expression is restricted to the liver and large vascular structures, but fibronectin expression increases dramatically in injured lungs. Since animals lacking fibronectin or its integrin receptors are embryonic lethal, emphasizing their importance in embryogenesis [30], the exact role of fibronectin in adult lungs is unclear. However, much evidence collected *in vitro* has implicated fibronectin in the reepithelialization of denuded basement membranes, the deposition of collagen, the proliferation of fibroblasts, the control of infection through its nonopsonin immune activity, and the recruitment of immune cells into tissues due to the chemotactic activity of some fibronectin fragments [31, 32]. These functions have led investigators to suggest that fibronectin expression is necessary for tissue repair after injury. On the other hand, excessive fibronectin deposition is considered detrimental since it might lead to exaggerated tissue repair responses that result in permanent tissue damage. The observation that animals deficient in fibronectin EDA are protected in experimental lung injury induced by bleomycin supports this idea [33, 34]. More interesting is the finding that regulated splicing of the fibronectin EDA exon is essential for normal lifespan in mice [34].

As stated before, alterations in the expression of collagens have been documented in senescent lungs [35]. In the same study, Sueblinvong et al. found changes in the mRNA expression of fibrillar collagens type I and III [14]. Interestingly, these changes were associated with increased gelatinolytic activity and expression of the MMPs MMP-2 and MMP-9. These gelatinases have been shown to degrade collagens, among other matrices, and have been implicated in many inflammatory and fibrotic lung disorders such as acute lung injury [36]. This is interesting considering that in a population-based study of 888 subjects aged 70 years, researchers found that impaired lung function in the elderly was associated with alterations in the serum levels of MMP-9, TIMP-1 (for tissue inhibitor of metalloproteinase-1), and their ratio. Specifically, lower force expiratory volume in the first second (FEV$_1$) was associated significantly (although weakly) with higher serum levels of MMP-9 and TIMP-1. The authors concluded that impaired lung function might be related to extracellular matrix remodeling; this association was stronger in men than in women [37]. It should be noted, however, that the significant percent of subjects with smoking history included in this study might have influenced the results. Using immunohistochemistry, others reported increased levels of TIMP-3 (another MMP inhibitor) in lung, kidney, and eye with aging [38]. In other work, researchers reported increased peribronchial collagen type I and II, while MMP-1 and MMP-2 activity dropped and MMP-9 levels were slightly decreased. These changes were associated with elevations in TIMP-1 and TIMP-2 [35].

The studies described earlier reveal inconsistencies in the data collected. However, when considered together, they suggest that aging is indeed associated with alterations in the expression and deposition of collagens and other matrices. Matrix degradation also appears affected due to changes in the expression and activity of matrix-degrading proteases (i.e., MMPs) and their inhibitors (i.e., TIMPs). Recently, investigations have pointed to granzymes as yet another group of proteases capable of matrix degradation in diseases of aging [39]. Granzyme B, one of the five human granzymes, is a serine protease capable of promoting apoptosis that has extracellular proteolytic activity [39]. Together with granzyme A, granzyme B has been implicated in immune-mediated killing of transformed cells or cells that are allogeneic or infected with viruses. Granzymes have also been implicated in the pathogenesis of disorders related to chronic inflammation such as chronic obstructive pulmonary disease and atherosclerosis [39, 40]. Granzymes act by entering the cell and promoting apoptosis through caspase-dependent and caspase-independent mechanisms, but they also have protein-degrading activity extracellularly [40]. They are produced by many pulmonary cells including bronchial and alveolar epithelial cells as well as alveolar macrophages. Granzyme B is also produced by mast cells, B and T cells, and chondrocytes. In a model of skin aging using apolipoprotein E (ApoE) knockout mice, ApoE/granzyme B double-knockout mice were fed a high-fat diet for 30 weeks. ApoE knockout mice showed hair graying, hair loss, skin thinning, loss of collagen density, and increased skin pathologies including collagen remodeling and reduced decorin compared to wild-type controls. These changes were more prominent in ApoE knockout animals fed with high-fat diets. However, the ApoE/granzyme B double knockouts showed protection against skin thinning, matrix degradation, and loss of dermal collagen density, suggesting a role for granzyme B in the matrix changes linked to aging in this model [41]. In general, it is thought that oxidant stress, inflammation, and the recruitment of inflammatory cells into tissues promote the release of granzymes which, in turn, engage in tissue damage through extracellular matrix fragmentation and cellular apoptosis [39].

It appears that aging lungs not only present with alterations in the content of connective tissue matrices, but the matrices themselves are altered. Aging is associated with protein modifications induced by the reaction of proteins with reducing sugars, resulting in the production of advanced glycation end products (AGEs) [42]. AGEs increase in aged tissues including the lung. Because of their longevity, collagens are characterized by AGE modifications through glycation, glycoxidation, and cross-links, and these changes may influence immune and tumor cell migration, among other processes [43, 44].

Together, these observations reveal that lung aging is associated with alterations in extracellular matrix expression/deposition, degradation, and protein modification. Importantly, most of the molecules described earlier have been linked to lung disrepair. For example, studies performed in animals with null mutations in TGF-β, fibronectin EDA, and MMPs (or abnormalities leading to poor activation of these molecules) have been implicated in experimental lung injury [25, 45, 46]. Consequently, one would assume that alterations in these genes would be associated with significant lung abnormalities. Paradoxically, this is not the case since senescent lungs look very much like young lungs when examined grossly and under light microscopy. Thus, counterregulatory mechanisms must be activated to ensure the integrity of the senescent lung. Further work to identify these counterregulatory mechanisms and their actions should lead to important new information and the identification of new targets for intervention.

11.4 The aging lung fibroblast

The increased expression of TGF-β and fibronectin EDA together with alterations noted in the production of collagens and matrix-degrading proteinases unveils a **profibrotic** phenotype in senescent lungs. The factors leading to this phenotypic change are unknown, but fibroblasts are likely involved. It is well known that the proliferative and migratory capabilities of human lung and skin fibroblasts decrease with aging [47]. Furthermore, aging is associated with a decline in the capacity for regeneration after pneumonectomy in most mammals. This has been linked to a loss of fibroblast clonogenicity and progressive myofibroblastic differentiation in murine lungs [48]. Studies examining the expression of Thy-1 on lung fibroblasts also suggest a change in phenotype. This glycosylphosphatidylinositol-linked cell surface glycoprotein regulates myofibroblast transdifferentiation and fibroblast survival [49]. Thy-1 mRNA expression was shown to be decreased in fibroblasts harvested from senescent lungs when compared to those isolated from the lungs of young animals [14]. This is interesting because there is decreased Thy-1 expression in patients with fibrosing lung disease [50, 51]. Further work designed to evaluate the phenotypic changes in fibroblasts and other lung cells that take place during aging and how they impact connective tissue composition, lung mechanics, and cellular functions after injury should be considered in order to determine the true role of these cells in lung aging.

11.5 Potential role of oxidant stress in triggering remodeling in the aging lung

Considering the aforementioned information, it is logical to wonder about the factors that trigger tissue remodeling in the aging lung. Inflammation has long been considered a factor capable of initiating and sustaining lung remodeling [52]. More recently, attention has turned to oxidant stress, and all evidence available to date indicates that aging is associated with oxidant stress [53]. Also, there is evidence in favor of the role for oxidant stress in lung disease [54, 55, 56, 57, 58, 59, 60, 61]. Some of these alterations are related to inefficient production of antioxidants. For example, during endotoxemia, the downregulation of extracellular superoxide dismutase is more profound and prolonged in aged mice compared to young mice [62]. Thus, it is reasonable to postulate that chronic oxidant stress elicits tissue remodeling in the aging lung. Although this remains to be confirmed experimentally, oxidants have been shown to stimulate the expression of several matrix components (e.g., fibronectin), proinflammatory transcription factors, as well as the expression and activation of growth factors and MMPs [63]. However, failure to affect outcomes in subjects with lung disease via the use of antioxidants has led many to postulate that the development and maintenance of oxidant stress in humans is complex and represents much more than an imbalance between oxidants and antioxidants.

Oxidant stress can be generated through a number of mechanisms ranging from aberrant function of NADPH oxidase, alterations in glutathione transport, and the generation of reactive oxygen and nitrogen species, among other mechanisms. Another mechanism of oxidative stress is related to the oxidation of thiol/disulfide pairs such as cysteine and cystine (Cys/CySS), glutathione and glutathione disulfide (GSH/GSSG), and thioredoxin, which results in alterations in their redox potential, hereafter termed **Eh**. An intriguing aspect of the Eh Cys/CySS is that it is operative extracellularly, while the Eh GSH/GSSG and thioredoxin predominate intracellularly (with thioredoxin predominating in the mitochondria) [64]. Jones and others have documented the oxidation of the Eh Cys/CySS in aging [65, 66] and have suggested that these changes in the redox potential of extracellular thiol pairs serve as independent transducers of oxidant stress. The novelty of this signaling mechanism is that it can trigger intracellular redox signaling by oxidation of membrane-bound proteins, and this oxidation can trigger the downstream generation of reactive oxygen species (ROS).

The mechanisms by which the oxidation of Eh Cys/CySS triggers intracellular signal transduction are unclear. Dean P. Jones has documented that the steady-state Eh Cys/CySS within cells can be sufficiently oxidized (>90 mV) to function as an oxidant in redox signaling.

Thus, as he points out, "the Cys/CySS redox couple represents a newly recognized node in the circuitry for biologic redox signaling and control" [66]. However, how exactly alterations in redox potential influence cell functions is unclear. Jones and his colleagues showed the activation of the p44/p42 MAPK pathway by extracellular Eh Cys/CySS in Caco-2 cells, and this was associated with increased cell proliferation [67]. In endothelial cells, an oxidized Eh Cys/CySS was sufficient to activate cellular ROS generation. More recently, and relevant to the lung, Ramirez et al. demonstrated that an oxidized extracellular Eh Cys/CySS results in the activation of TGF-β_1 /Smad3 signaling in lung fibroblasts and that this was associated with the induction of fibronectin production, the expression of smooth muscle cell markers (i.e., α-smooth muscle actin), and increased cellular proliferation [68]. Others have begun to evaluate the redox potential found in animals exposed to bleomycin, a well-known model for lung injury. These studies revealed that bleomycin-induced lung injury further enhances the oxidation of the Cys/CySS and GSH/GSSG redox potentials [69]. In other work, it was shown that Eh Cys/CySS controls proinflammatory cytokine levels [70]. These are interesting observations considering that Cys availability and extracellular Eh Cys/CySS are dependent on age, nutrition, disease, and environmental exposures. Thus, the oxidation of the Eh Cys/CySS is likely to represent yet another pathway by which aging could exert its detrimental effects on extracellular matrix remodeling.

11.6 Implications for remodeling of the lung extracellular matrix in the aged lung

The studies presented previously indicate that the uninjured aged lung is influenced by oxidant stress and other factors capable of inducing lung tissue remodeling. These observations raise several questions: Is tissue remodeling in the aging lung similar to the tissue remodeling observed after lung injury? How does remodeling affect lung cell functions? What are the implications for the activation of tissue remodeling in the aging lung as they relate to injury and repair as well as susceptibility to chronic lung disease? The first question is an intriguing one considering that, despite the many alterations in the expression of tissue remodeling genes documented in senescent lungs, there are few, if any, changes observed in overall tissue architecture. In fact, and in contrast to the tissue remodeling observed after what we would traditionally term lung injury, the changes in matrix remodeling described in senescent lungs are only detectable after careful evaluation of lung tissue with immunohistochemistry and related techniques. We have termed this process **transitional remodeling**, a nondevelopmental process characterized by alterations in the expression of tissue remodeling genes, resulting in changes in the relative composition of the extracellular matrix in the absence of overt changes in tissue architecture. We believe that transitional remodeling is a predisease state that does not result in significant lung dysfunction by itself, but that predisposes the lung to exaggerated repair responses after injury, thereby rendering the host susceptible to a **second hit** that drives permanent damage.

It is important to emphasize that transitional remodeling has not been well documented in otherwise healthy elderly humans, which is not surprising considering that most of these subjects fail to show manifestations of lung disease. Also, the consequences of transitional remodeling in the senescent lung remain speculative. However, transitional remodeling is increasingly being identified in conditions associated with lung disease. In adult and perinatal murine lungs, nicotine exposure is associated with increased expression of fibronectin and type I collagen without overt changes in lung architecture [71, 72]. Careful morphometric analysis of lung tissue in the offspring of nicotine-exposed animals showed mild thickening of airways and increased branching, but the overall branching pattern of the developing lung appeared normal. When physiological studies were performed, however, airflow limitation and increased hyperreactivity in response to methacholine stimulation were identified [73]. Similar observations were made in nonhuman primates [74], and these changes may explain the increased incidence of obstructive airway diseases (e.g., asthma) shown in adults exposed to tobacco products during the perinatal period and early childhood [75]. Chronic alcohol exposure is also associated with transitional remodeling in both rodents and humans as highlighted by increased expression of TGF-β, fibronectin, and MMPs in lungs [76, 77, 78, 79]. As before, the lungs of alcohol-exposed animals seem normal at baseline, but they show increased injury when exposed to endotoxin. These and other observations have been used to define what is now known as the **alcoholic lung phenotype** and explain, at least in part, the increased incidence and mortality due to acute lung injury in chronic alcoholics in the setting of sepsis [80]. Others have reported increased gelatinolytic activity in the bronchoalveolar lavage fluid of transplanted lungs way before the seemingly **healthy** hosts developed manifestations of chronic rejection [81]. By manifesting alterations in matrix composition without overt changes in tissue architecture and yet showing increased susceptibility to injury, senescent lungs may represent another example of transitional remodeling.

These studies suggest that internal factors (e.g., aging), external or environmental factors (e.g., nicotine and ethanol), and immune-mediated processes (e.g., lung transplantation) can trigger transitional remodeling perhaps through induction of oxidant stress although this requires further exploration. However, if so, one can only assume that these are just few of the many settings and exposures in which transitional remodeling can be activated. Systemic illness, diet, and medications could potentially trigger this process. Although the exact consequences of transitional remodeling are unknown, it is likely that this represents a predisease state that renders the host susceptible to a second hit. If these ideas are confirmed, one would predict that biomarkers of transitional remodeling (e.g., detection of excess fibronectin or MMPs in bronchoalveolar lavage fluid) in otherwise healthy hosts might help identify subjects at increased susceptibility to disrepair after injury. Furthermore, if transitional remodeling is found to be pathogenetic, this process may turn out to be a reasonable target for intervention.

The second question is equally important: How does transitional remodeling affect the aging lung considering that it does not cause overt changes in tissue architecture and function? This question remains unanswered. However, matrix remodeling can only affect biological processes if cells have the capacity to recognize them. We know that this is indeed the case and studies *in vitro* have been very informative in this regard. Cells recognize extracellular matrices through surface receptors capable of signal transduction such as integrins. Integrins are a large family of heterodimeric transmembrane receptors capable of linking extracellular and intracellular signals that affect gene expression [82]. These receptors are made of α and β subunits joined by noncovalent bonds and expressed at the surface of all human cells. Because β subunits are often shared by several integrins with similar functions, while α subunits provide ligand specificity, integrins were initially classified into subfamilies based on their β subunit (e.g., β1, β2, β3), a classification that is still useful. The β1 integrin subfamily is especially relevant to this discussion considering that it contains many of the best characterized matrix-binding integrin receptors including receptors for collagens and laminin (α1β1, α2β1, and α3β1), fibronectin (α5β1), and laminin (α6β1) [83]. These and other integrins represent a critical tool in the repertoire of cells involved in matrix recognition.

Agents capable of inhibiting the expression of specific matrix components or blocking their binding to integrin receptors have been used to elucidate the functions of the extracellular matrix. Early studies showed that fibronectin and collagens serve to enhance cellular adhesion, migration, proliferation, apoptosis, and differentiation depending on the cell type [84, 85, 86]. Extracellular matrix fragments are known for their chemotactic activity, while gradients in the insoluble matrix can promote haptotaxis (the migration of cells toward an insoluble gradient) [87]. These *in vitro* studies have been extended greatly to test more complex processes. For example, it has been reported that different matrices influence the expression of cell–cell adhesion molecules in ways that affect the permeability of epithelial cell monolayers [88]. Extracellular matrices like fibronectin, type I collagen, and fibrin can activate immune cells and stimulate their production of proinflammatory cytokines through integrin-mediated activation of specific transcription factors [89]. These and other events are regulated by the induction of specific intracellular signaling pathways ranging from protein phosphorylation, the influx of ions, and the activation of distinct protein kinases (e.g., PKC, AKT, Erks) followed by differential gene expression [90]. These integrin-mediated events have further been linked to effects on embryogenesis, inflammation, cellular apoptosis, and carcinogenesis in lung [91]. Integrins also transmit mechanical forces from the environment (mechanotransduction). In doing so, tissue stiffness can affect cellular functions. As expected, increased measures of lung stiffness have been demonstrated in models of lung fibrosis [92], but it is not clear if transitional remodeling exerts sufficient tension on cells to affect function. Finally, we suspect that transitional remodeling not only affects the aforementioned processes but may also influence endoplasmic reticulum stress and stem cell recruitment to injured lungs which have been implicated in aging more recently [93, 94]. With regard to the latter, one can envision how changes in composition of the lung matrix during aging may affect the recruitment and differentiation of stem cells depending on the repertoire of integrins expressed on their surface.

Considering the ability of lung cells to recognize extracellular matrix composition, it stands to reason that alterations in matrix composition, even as subtle as those proposed to occur during transitional remodeling, would impact cellular behavior. Under some circumstances, these changes may promote adaptive repair after injury, while in others, maladaptive mechanisms would be introduced. We postulate that transitional remodeling in the aging lung is a maladaptive mechanism that leads to enhanced susceptibility to infection and injury, but this awaits further experimentation (Figure 11.1).

11.7 Conclusions

Lung senescence is associated with significant alterations in the expression, deposition, modification, and degradation of extracellular matrix proteins. This is due to the increased expression of growth factor genes, the release of matrix-degrading proteases, and the generation of AGEs, among other mechanisms. Chronic inflammation, oxidant stress, and environmental exposures in the elderly may affect the phenotype of fibroblasts and other lung cells, thereby promoting the release and, ultimately, the assembly of fibronectin, collagens, and other matrix components. These processes lead to subtle alterations in the relative composition of the matrix without causing overt changes in overall lung tissue architecture (transitional remodeling) (Figure 11.2). Through integrins and other surface receptors, distinct matrix components and alterations in tissue stiffness (which may activate signaling through mechanotransduction) may elicit intracellular signals that promote differential gene expression. Although these events do not appear to greatly affect lung function at baseline, they

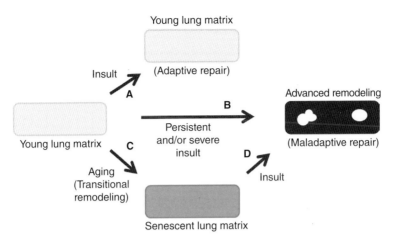

Figure 11.1 Hypothetical impact of transitional remodeling to tissue repair after injury. The young lung, with its natural extracellular matrix composition, is capable of repair after injury, leading to reestablishment of the original lung architecture and function after an insult, A (adaptive repair). However, severe and/or long-lasting insults may lead to permanent damage, B. During aging, the senescent lung undergoes transitional remodeling that alters the relative composition of the extracellular matrix, C. This does not have significant implications for lung structure and function at baseline. However, since lung resident and incoming cells interact with matrices through integrins and other receptors, these cells recognize this new matrix and, in response, engage in exuberant repair responses after an insult, leading to maladaptive repair and permanent tissue damage, D.

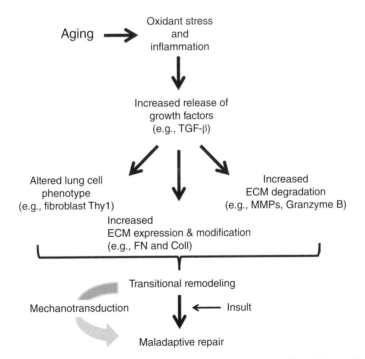

Figure 11.2 Extracellular matrix remodeling in senescent lungs. It is speculated that oxidant stress and chronic inflammation trigger the expression of growth factors and other mediators capable of activating tissue remodeling in the senescent lung. This results in altered phenotype in lung fibroblasts and perhaps other cells as well as increased expression and degradation of matrix proteins, leading to changes in the relative composition of the extracellular matrix. At baseline, this transitional remodeling may not cause major changes to tissue function. However, they may render the host susceptible to disrepair after injury, leading to permanent tissue damage.

might significantly influence immune and reparative mechanisms of action triggered by an insult to the lung, thereby leading to maladaptive disrepair and culminating in permanent tissue damage. The true role of transitional remodeling in lung aging is unknown. However, a better understanding of the counterregulatory mechanisms that maintain the integrity of the senescent lung despite the changes described earlier would likely lead to important insights. Furthermore, the identification of potential targets for intervention in transitional remodeling may accelerate the development of effective preventive or treatment strategies. In this new age of genetic screening and manipulation, lung senescence is considered an important focus of investigation and new discoveries in this field are expected to decelerate the aging-related changes in extracellular matrix composition, hopefully decreasing susceptibility to disrepair and sustaining lung health.

Acknowledgments

The author thanks Jeffrey Ritzenthaler, Rafael L. Perez, and Edilson Torres for helpful comments on the manuscript. This work was supported by grant AG038652 from the National Institute on Aging, National Institutes of Health.

References

1. Poljsak B, Dahmane RG, Godic A. Intrinsic skin aging: the role of oxidative stress. *Acta Dermatovenerol Alp Panonica Adriat* 2012;21:33–36.
2. Janssens JP, Pache JC, Nicod LP. Physiological changes in respiratory function associated with ageing. *Eur Respir J* 1999;13:197–205.
3. Faner R, Rojas M, MacNee W, Agusti A. Abnormal lung aging in chronic obstructive pulmonary disease and idiopathic pulmonary fibrosis. *Am J Respir Crit Care Med* 2012;186:306–313.
4. Siner JM, Pisani MA. Mechanical ventilation and acute respiratory distress syndrome in older patients. *Clin Chest Med* 2007;28:783–791.
5. Ware LB. Prognostic determinants of acute respiratory distress syndrome in adults: impact on clinical trial design. *Crit Care Med* 2005;33:S217–S222.
6. Hinojosa E, Boyd AR, Orihuela CJ. Age-associated inflammation and toll-like receptor dysfunction prime the lungs for pneumococcal pneumonia. *J Infect Dis* 2009;200:546–554.
7. Vincent R, Vu D, Hatch G, et al. Sensitivity of lungs of aging Fischer 344 rats to ozone: assessment by bronchoalveolar lavage. *Am J Physiol* 1996;271:L555–L565.
8. Ito Y, Betsuyaku T, Nasuhara Y, Nishimura M. Lipopolysaccharide-induced neutrophilic inflammation in the lungs differs with age. *Exp Lung Res* 2007;33:375–384.
9. Darniot M, Pitoiset C, Petrella T, Aho S, Pothier P, Manoha C. Age-associated aggravation of clinical disease after primary metapneumovirus infection of BALB/c mice. *J Virol* 2009;83:3323–3332.
10. Nin N, Lorente JA, De Paula M, et al. Aging increases the susceptibility to injurious mechanical ventilation. *Intensive Care Med* 2008;34:923–931.
11. Matulionis DH. Chronic cigarette smoke inhalation and aging in mice: 1. Morphologic and functional lung abnormalities. *Exp Lung Res* 1984;7:237–256.
12. Franceschi C, Bonafe M, Valensin S. Human immunosenescence: the prevailing of innate immunity, the failing of clonotypic immunity, and the filling of immunological space. *Vaccine* 2000;18:1717–1720.
13. Gomez CR, Hirano S, Cutro BT, et al. Advanced age exacerbates the pulmonary inflammatory response after lipopolysaccharide exposure. *Crit Care Med* 2007;35:246–251.
14. Sueblinvong V, Neujahr DC, Mills ST, et al. Predisposition for disrepair in the aged lung. *Am J Med Sci* 2012;344:41–51.
15. Meyer KC. Aging. *Proc Am Thorac Soc* 2005;2:433–439.
16. Kretlow JD, Jin YQ, Liu W, et al. Donor age and cell passage affects differentiation potential of murine bone marrow-derived stem cells. *BMC Cell Biol* 2008;9:60–65.

17. Finley LW, Haigis MC. The coordination of nuclear and mitochondrial communication during aging and calorie restriction. *Ageing Res Rev* 2009;8:173–188.

18. Blagosklonny MV. Revisiting the antagonistic pleiotropy theory of aging: TOR-driven program and quasi-program. *Cell Cycle* 2010;9:3151–3156.

19. Estep PW 3rd, Warner JB, Bulyk ML. Short-term calorie restriction in male mice feminizes gene expression and alters key regulators of conserved aging regulatory pathways. *PLoS One* 2009;4(4):e5242.

20. Ren W-Y, Li L, Zhao R-Y, Zhu L. Age associated changes in pulmonary function: a comparison of pulmonary function parameters in healthy young adults and the elderly living in Shanghai. *Chin Med J* 2012;125:3064–3068.

21. D'Errico A, Scarani P, Colosimo E, Spina M, Grigioni WF, Mancini AM. Changes in the alveolar connective tissue of the ageing lung. An immunohistochemical study. *Virchows Arch A Pathol Anat Histopathol* 1989;15:137–144.

22. Andreotti L, Bussotti A, Cammelli D, Aiello E, Sampognaro S. Connective tissue in aging lung. *Gerontology* 1983;29:377–387.

23. Huang K, Rabold R, Schofield B, Mitzner W, Tankersley CG. Age-dependent changes of airway and lung parenchyma in C57BL/6J mice. *J Appl Physiol* 2007;102:200–206.

24. Takubo Y, Hirai T, Muro S, Kogishi K, Hosokawa M, Mishima M. Age-associated changes in elastin and collagen content and the proportion of types I and III changes in the lungs of mice. *Exp Gerontol* 1999;34:353–364.

25. Dhainaut JF, Charpentier J, Chiche JD. Transforming growth factor-beta: a mediator of cell regulation in acute respiratory distress syndrome. *Crit Care Med* 2003;31:S258–S264.

26. Pittet JF, Griffiths MJ, Geiser T, et al. TGF-beta is a critical mediator of acute lung injury. *J Clin Invest* 2001;107:1537–1544.

27. Sheppard D. Transforming growth factor beta: a central modulator of pulmonary and airway inflammation and fibrosis. *Proc Am Thorac Soc* 2006;3:413–417.

28. Ramirez AM, Takagawa S, Sekosan M, Jaffe A, Varga J, Roman J. Smad3 deficiency ameliorates experimental obliterative bronchiolitis in a heterotopic tracheal transplantation model. *Am J Pathol* 2004;165:1223–1232.

29. Roman J. Fibronectin and fibronectin receptors in lung development. *Exp Lung Res* 1997;23:147–159.

30. Georges-Labouesse EN, George EL, Rayburn H, Hynes RO. Mesodermal development in mouse embryos mutant for fibronectin. *Dev Dyn* 1996;207:145–156.

31. Limper AH, Roman J. Fibronectin: a versatile matrix protein with roles in thoracic development, repair and infection. *Chest* 1992;101:163–173.

32. Larsen M, Artzym VV, Green JA, Yamada KM. The matrix reorganized: extracellular matrix remodeling and integrin signaling. *Curr Opin Cell Biol* 2006;18:463–471.

33. White ES, Baralle FE, Muro AF. New insights into form and function of fibronectin splice variants. *J Pathol* 2008;216:1–14.

34. Muro AF, Chauhan AK, Gajovic S, et al. Regulated splicing of the fibronectin EDA exon is essential for proper skin wound healing and normal lifespan. *J Cell Biol* 2003;162:149–160.

35. Calabresi C, Arosio B, Galimberti L, et al. Natural aging, expression of fibrosis-related genes and collagen deposition in rat lung. *Exp Gerontol* 2007;42:1003–1011.

36. Davey A, McAuley DF, O'Kane CM. Matrix metalloproteinases in acute lung injury: mediators of injury and drivers of repair. *Eur Respir J* 2011;38:959–970.

37. Olafsdottir IS, Janson C, Lind L, Hulthe J, Gunnbjornsdottir M, Sundstrom J. Serum levels of matrix metalloproteinase-9, tissue inhibitors of metalloproteinase-1 and their ratio are associated with impaired lung function in the elderly: a population-based study. *Respirology* 2010;15:530–535.

38. Macgregor AM, Eberhart CG, Fraig M, Lu J, Halushka MK. Tissue inhibitor of matrix metalloproteinase-3 levels in the extracellular matrix of lung, kidney, and eye increase with age. *J Histochem Cytochem* 2009;57:207–213.

39. Hendel A, Hiebert PR, Boivin WA, Williams SJ, Granville DJ. Granzymes in age-related cardiovascular and pulmonary diseases. *Cell Death Differ* 2010;17:596–606.

40. Boivin WA, Cooper DM, Hiebert PR, Granville DJ. Intracellular versus extracellular granzyme B in immunity and disease: challenging the dogma. *Lab Invest* 2009;89:1195–1220.

41. Hiebert PR, Boivin WA, Abraham T, Pazooki S, Zhao H, Granville DJ. Granzyme B contributes to extracellular matrix remodeling and skin aging in apolipoprotein E knockout mice. *Exp Gerontol* 2011;46:489–499.

42. Bellmunt MJ, Portero M, Pamplona R, Muntaner M, Prat J. Age-related fluorescence in rat lung collagen. *Lung* 1995;173:177–185.

43. Sell DR, Nelson JF, Monnier VM. Effect of chronic aminoguanidine treatment on age-related glycation, glycoxidation, and collagen cross-linking in the Fisher 344 rat. *J Gerontol A Biol Sci Med Sci* 2011;56:B405–B411.

44. Bartling B, Desole M, Rohrbach S, Silber R-E, Simm A. Age-associated changes of extracellular matrix collagen impair lung cancer cell migration. *FASEB J* 2009;23:1510–1520.

45. Muro AF, Moretti FA, Moore BB, et al. An essential role for fibronectin extra type III domain A in pulmonary fibrosis. *Am J Respir Crit Care Med* 2007;177:638–645.

46. Albaiceta GM, Gutierrez-Fernandez A, Parra D, et al. Lack of MMP-9 worsens ventilator-induced lung injury. *Am J Physiol* 2008;294:L535–L543.

47. Kondo H, Yonezawa Y. Changes in the migratory ability of human lung and skin fibroblasts during in vitro aging and in vivo cellular senescence. *Mech Ageing Dev* 1992;63:223–233.

48. Paxson JA, Gruntman A, Parkin CD, et al. Age-dependent decline in mouse lung regeneration with loss of lung fibroblast clonogenicity and increased myofibroblastic differentiation. *PLoS One* 2011;6(8):e23232.

49. Zhou Y, Hagood JS, Murphy-Ullrich JE. Thy-1 expression regulates the ability of rat lung fibroblasts to activate transforming growth factor-beta in response to fibrogenic stimuli. *Am J Pathol* 2004;165: 659–669.

50. Hagood JS, Prabhakaran P, Kumbla P, et al. Loss of fibroblast Thy-1 expression correlates with lung fibrogenesis. *Am J Pathol* 2005;167:365–379.

51. Sanders YY, Pardo A, Selman M, et al. Thy-1 promoter hypermethylation: a novel epigenetic pathogenic mechanism in pulmonary fibrosis. *Am J Respir Cell Mol Biol* 2008;39:610–618.

52. dos Santos G, Kutuzov MA, Ridge KM. The inflammasome in lung disease. *Am J Physiol Lung Cell Mol Physiol* 2012;303:L627–L633.

53. Harman D. Free radical theory of aging: an update: increasing the functional life span. *Ann NY Acad Sci* 2006;1067:10–21.

54. Rahman I, Biswas SK, Kode A. Oxidant and antioxidant balance in the airways and airway diseases. *Eur J Pharmacol* 2006;533:222–239.

55. Rahman I. Regulation of glutathione in inflammation and chronic lung disease. *Mutat Res* 2005;579:58–80.

56. Bellocq A, Azoulay E, Marullo S, et al. Reactive oxygen and nitrogen intermediates increase transforming growth factor beta 1 release from human epithelial alveolar cells through two different mechanisms. *Am J Respir Crit Care Med* 1999;21:128–136.

57. Comhair SA, Erzurum SC. Antioxidant responses to oxidant-mediated lung diseases. *Am J Physiol* 2002;283:L246–L255.

58. Tasaka S, Amaya F, Hashimoto S, Ishizaka A. Roles of oxidants and redox signaling in the pathogenesis of acute respiratory distress syndrome. *Antioxid Redox Signal* 2008;10:739–753.

59. Lang JD, McArdle PJ, O'Reilly PJ, Matalon S. Oxidant-antioxidant balance in acute lung injury. *Chest* 2002;122:314S–320S.

60. Guo RF, Ward PA. Role of oxidants in lung injury during sepsis. *Antioxid Redox Signal* 2007;9:1991–2002.

61. Biswas SK, Rahman I. Environmental toxicity, redox signaling, and lung inflammation: the role of glutathione. *Mol Aspects Med* 2009;30:60–76.

62. Starr ME, Ueda J, Yamamoto S, Evers BM, Saito H. The effects of aging on pulmonary oxidative damage, protein nitration, and extracellular superoxide dismutase down-regulation during systemic inflammation. *Free Radic Biol Med* 2011;50:371–380.

63. Bargagli E, Olivieri C, Bennett D, Prasse A, Muller-Quernheim J, Rottoli P. Oxidative stress in the pathogenesis of diffuse lung diseases: a review. *Respir Med* 2009;103:1245–1256.

64. Jones DP. Extracellular redox state: refining the definition of oxidative stress in aging. *Rejuvenation Res* 2006;9:169–181.

65. Jones DP. Redefining oxidative stress. *Antioxid Redox Signal* 2006;8:1865–1879.

66. Jones DP, Go YM, Anderson CL, Ziegler TR, Kinkade JM Jr, Kirlin WG. Cysteine/cystine couple is a newly recognized node in the circuitry for biologic redox signaling and control. *FASEB J* 2004;28:1246–1248.

67. Nkabyo YS, Go YM, Ziegler TR, Jones DP. Extracellular cysteine/cystine redox regulates the p44/p42 MAPK pathway by metalloproteinase-dependent epidermal growth factor receptor signaling. *Am J Physiol Gastrointest Liver Physiol* 2005;289:G70–G78.

68. Ramirez A, Ramadan B, Ritzenthaler JD, Rivera HN, Jones DP, Roman J. Extracellular cysteine/cystine redox potential controls lung fibroblast proliferation and matrix expression through upregulation of transforming growth factor-beta. *Am J Physiol* 2007;293:L972–L981.
69. Iyer SS, Ramirez AM, Ritzenthaler JD, et al. Oxidation of extracellular cysteine/cystine redox state in bleomycin-induced lung fibrosis. *Am J Physiol* 2008;296:L37–L45.
70. Iyer SS, Accardi CJ, Ziegler TR, et al. Cysteine redox potential determines pro-inflammatory IL-1β levels. *PLoS One* 2009;4(3):e5017.
71. Roman J, Ritzenthaler JD, Gil-Acosta A, Rivera HN, Roser-Page S. Nicotine and fibronectin expression in lung fibroblasts: implications for tobacco-related lung tissue remodeling. *FASEB J* 2004;18:1436–1438.
72. Wongtrakool C, Roser-Page S, Rivera HN, Roman J. Nicotine alters lung branching morphogenesis through the α_7 nicotinic acetylcholine receptor. *Am J Physiol Lung Cell Mol Physiol* 2007;293:L611–L618.
73. Wongtrakool C, Wang N, Hyde DM, Roman J, Spindel ER. Prenatal nicotine exposure alters lung function and airway geometry through α_7 nicotinic receptors. *Am J Respir Cell Mol Biol* 2012;46:695–702.
74. Sekhon HS, Jia Y, Raab R, et al. Prenatal nicotine increases pulmonary α7 nicotinic receptor expression and alters fetal lung development in monkeys. *J Clin Invest* 1999;103:637–647.
75. Cheraghi M, Salvi S. Environmental tobacco smoke (ETS) and respiratory health in children. *Eur J Pediatr* 2009;168:897–905.
76. Roman J, Ritzenthaler JD, Bechara R, Brown LA, Guidot D. Ethanol stimulates the expression of fibronectin in lung fibroblasts via kinase-dependent signals that activate CREB. *Am J Physiol* 2005;288:L975–L987.
77. Brown LA, Ritzenthaler JD, Guidot DM, Roman J. Alveolar type II cells from ethanol-fed rats produce a fibronectin-enriched extracellular matrix that promotes monocyte activation. *Alcohol* 2007;41:317–324.
78. Bechara RI, Brown LA, Roman J, Joshi PC, Guidot DM. Transforming growth factor b1 expression and activation is increased in the alcoholic rat lung. *Am J Respir Crit Care Med* 2004;170:188–194.
79. Louis M, Brown LA, Moss IM, Roman J, Guidot DM. Ethanol ingestion increases activation of matrix metalloproteinases in rat lungs during acute endotoxemia. *Am J Respir Crit Care Med* 1999;160:1354–1360.
80. Guidot DM, Roman J. Chronic ethanol ingestion increases susceptibility to acute lung injury: role of oxidant stress and tissue remodeling. *Chest* 2002;122:309S–314S.
81. Ramirez AM, Nunley DR, Rojas M, Roman J. Activation of tissue remodeling precedes obliterative bronchiolitis in lung transplant recipients. *Biomark Insights* 2008;3:351–359.
82. Humphries MJ, Campbell ID. Integrin structure, activation, and interactions. *Cold Spring Harb Perspect Biol* 2011;3:a004994.
83. Pinon P, Wehrle-Haller B. Integrins: versatile receptors controlling melanocyte adhesion, migration and proliferation. *Pigment Cell Melanoma Res* 2011;24:282–294.
84. Clark RA, Lanigan JM, DellaPelle P, Manseau E, Dvorak HF, Colvin RB. Fibronectin and fibrin provide a provisional matrix for epidermal cell migration during wound reepithelialization. *J Invest Dermatol* 1982;79:264–269.
85. Bitterman PB, Rennard SI, Adelberg S, Crystal RG. Role of fibronectin as a growth factor for fibroblasts. *J Cell Biol* 1983;97:1925–1932.
86. Zhang Y, Wang H. Integrin signaling and function in immune cells. *Immunology* 2012;135:268–275.
87. Roman J. Extracellular matrix and lung inflammation. *Immunol Res* 1996;15:163–178.
88. Koval M, Ward C, Findley MK, Roser-Page S, Helms MN, Roman J. Extracellular matrix influences alveolar epithelial claudin expression and barrier function. *Am J Respir Cell Mol Biol* 2010;42:172–180.
89. Graves KL, Roman J. Fibronectin modulates expression of interleukin-1b and its receptors antagonist in human mononuclear cells. *Am J Physiol* 1996;271:L61–L69.
90. Kinashi T. Overview of integrin signaling in the immune system. *Methods Mol Biol* 2012:757:261–278.
91. Roman J, Ritzenthaler JD, Roser-Page S, Sun X, Han S. α5β1 integrin expression is essential for tumor progression in experimental lung cancer. *Am J Respir Cell Mol Biol* 2010;43:684–691.
92. Liu F, Mih JD, Shea BS, et al. Feedback amplification of fibrosis through matrix stiffening and COX-2 suppression. *J Cell Biol* 2010;190:693–706.
93. Torres-González E, Bueno M, Tanaka A, et al. Role of endoplasmic reticulum stress in age-related susceptibility to lung fibrosis. *Am J Respir Cell Mol Biol* 2012;46:748–756.
94. Liu J, Cao L, Finkel T. Oxidants, metabolism, and stem cell biology. *Free Radic Biol Med* 2011;51:2158–2162.

12 Aging Mesenchymal Stem Cells in Lung Disease

Maria G. Kapetanaki[1,2], Ana L. Mora[2], and Mauricio Rojas[1,2,3]

[1]Dorothy P. and Richard P. Simmons Center for Interstitial Lung Diseases, University of Pittsburgh School of Medicine, Pittsburgh, Pennsylvania, USA
[2]Division of Pulmonary, Allergy and Critical Care Medicine, University of Pittsburgh School of Medicine, Pittsburgh, Pennsylvania, USA
[3]McGowan Institute for Regenerative Medicine, University of Pittsburgh School of Medicine, Pittsburgh, Pennsylvania, USA

12.1 Aging and lung diseases

Epidemiological studies indicate that aging is associated with an increased incidence of a variety of chronic lung diseases affecting a significant portion of the population [1, 2, 3, 4, 5]. The prevalence of chronic obstructive pulmonary disease (COPD) increases with age for both men and women throughout most of the lifespan, with the highest prevalence among individuals aged 65–84 [6, 7]. In 2005, approximately 1 in 20 deaths in the United States had COPD as the underlying cause [5]. Similarly, the occurrence of idiopathic pulmonary fibrosis (IPF) increases in both prevalence and incidence in the sixth decade of life. Symptoms typically occur at the age of 50 to 70 years, and most patients are greater than60 years of age at the time of clinical presentation [2, 8]. Notably, both of the aforementioned diseases show an increased deposition of collagen and fibrotic tissue albeit in different locations of sometimes the same lung, suggesting common, at least in part, molecular mechanisms of pathogenesis. Currently, only limited therapeutic strategies exist to treat chronic lung diseases, most of which are symptomatic instead of causal.

Although not a chronic lung disease, acute respiratory distress syndrome (ARDS) is a very common clinical entity and a major cause of morbidity and mortality in the critical care setting especially for the elderly patients. Only in the US, 190,600 new cases of ARDS occur every year [9], and recent studies show an incidence of 306 per 100,000 person-years for people from 75 to 84 years old, compared to 16 cases per 100,000 person-years in teenagers between 15 and 19 years old. Mortality increases from 24% to 60% for those older than 85 years compared to 15–19-year-old patients. Furthermore, younger patients had a better rate of recovery than older patients, but neither group returned to previous level of functionality [10, 11, 12]. Although existing therapeutic approaches are contributing to the decrease in mortality, they, in many instances, aggravate lung injury.

Even if our understanding of the biology of aging has advanced remarkably, the molecular mechanisms linking aging to susceptibility to lung injury and disease remain unclear. Cellular senescence, oxidative stress, abnormal shortening of telomeres, apoptosis, and

Molecular Aspects of Aging: Understanding Lung Aging, First Edition. Edited by Mauricio Rojas, Silke Meiners and Claude Jourdan Le Saux.

epigenetic changes affecting gene expression have been proposed to contribute to the aging process and aging-associated diseases [13, 14, 15, 16]. Animal studies also support the link between aging and susceptibility to diseases by demonstrating an increased vulnerability of the aged lung to injury. Our group was the first to demonstrate that a single episode of lung injury by bleomycin [2, 17] or infection with gammaherpesvirus [18] in the lung causes severe progressive pulmonary fibrosis only in naturally aged wild-type mice, while the same level of injury can be achieved only in genetically modified young mice that are predisposed to lung injury. These observations, in some way, challenge the traditional concept that progressive fibrosis is the result of chronic injury. While all cells suffer upon age-related DNA damage, proliferating cells are infinitely more vulnerable as they undergo intense cycles of DNA replication. In many cases, DNA damage in highly proliferating cells is counteracted by apoptosis or cellular senescence to avoid the potential propagation of harmful mutations. This phenomenon may play a very important role in the occurrence of abnormal wound healing and fibrosis in the lung as it affects the proliferation of both lung and circulating progenitor cells that are involved in the repair process [1, 2, 4, 19]. From a multipotent mesenchymal stromal cell (MSC) perspective, aging can be defined as the "sum of primary restrictions in regenerative mechanisms of multicellular organisms" [20], and therefore, the study of the age-related changes that arise in MSCs is crucial, particularly for exploring their full therapeutic potential.

12.2 Mesenchymal stem cells (MSCs)

12.2.1 Description of MSCs

MSCs are multipotent cells that can differentiate into osteoblasts, chondrocytes, and adipocytes *in vitro* [21]. As all stem cells, they go into asymmetric differentiation and are capable of self-renewing, qualities that ensure the replenishing of the stem cell pool and the generation of new populations of cell-specific progenitor cells [22]. Although MSCs have been isolated from the connective tissue of almost every organ of the human body, this chapter will focus on bone marrow-derived MSCs (B-MSCs) as they are the well characterized and have high potential to be used in cell-based therapies for lung diseases.

12.2.2 Characterization of MSCs

B-MSCs are postembryonic stem cells which can differentiate into connective tissue of non-hematopoietic lineage. The accurate characterization of B-MSCs has been a complicated issue since there are no specific cell surface markers and they are found in extremely low frequency in the bone marrow. Enrichment of B-MSCs from crude bone marrow suspensions is achieved by selection for a plastic-adherent population that expresses neither hematopoietic nor endothelial cell surface markers but is positive for the expression of adhesion and stromal markers [23]. In the lack of a defined panel of unambiguous markers distinguishing B-MSCs, a criterion for establishing B-MSC phenotype is to use adherent cells isolated by cell sorting that (i) express CD44, CD73, CD90, and CD105; (ii) lack the expression of hematopoietic markers like CD45, CD34, and CD31; and finally (iii), in a trilineage differentiation assay, confirm their plasticity by the ability of the cells to differentiate into adipocytes, osteocytes, and chondrocytes [24].

12.2.3 Functional properties of MSCs

In vivo, B-MSCs are likely to participate in the regulation of hematopoiesis through their mature progeny (osteoblasts, adipocytes, and fibroblastic reticular cells). In recent years, B-MSCs have gained a lot of attention due to their immunomodulatory potential [25]. Although the big majority of the data is collected from *in vitro* studies and has the risk of leading to conclusions that deviate from what really happens *in vivo*, the current doctrine is that B-MSCs' interaction with innate and adaptive immunity mechanisms could establish a complex immunoregulatory network[1, 26, 27]. In summary, B-MSCs can inhibit the maturation of monocytes and hematopoietic progenitor cells into dendritic cells (DCs) as well as decrease the antigen-presenting ability and proinflammatory potential of mature DCs [28]. They can also inhibit the proliferation, interferon gamma (IFN gamma) production, and cytotoxic activity of both resting and preactivated natural killer (NK) cells, although that later interaction seems to be more complicated as it depends on the levels of the IFN gamma in the medium [29]. Finally, B-MSCs can reduce the respiratory burst of activated neutrophils. In adaptive immunity, B-MSCs are shown to inhibit antigen-specific T-cell proliferation and cytotoxicity and promote the generation of regulatory T cells [30, 31]. Interestingly, pathogens can interact with Toll-like receptors that express on B-MSC surface and activate a series of events that block the B-MSCs' inhibitory effect on T-cell proliferation. This flexible mechanism would protect the organism from becoming defenseless to infections by an excessive inhibition of T cells. Despite the large number of publications that describe the immunomodulatory effects of B-MSCs, there is still a lot of uncertainty regarding the mechanisms that are implicated to achieve these effects. It is generally accepted that cell-to-cell contact and secretion of soluble factors are two main ways how B-MSCs interact with their target cells. Examples of these factors are nitric oxide, indoleamine 2,3-dioxygenase (IDO), transforming growth factor-beta1 (TGFbeta1), hepatocyte growth factor (HGF), interleukin-10 (IL-10), prostaglandin E2 (PGE2), heme oxygenase-1 (HO1), interleukin-6 (IL-6), and soluble human leukocyte antigen-G5 (HLA-G5) [32–36].

It would be unjust to describe the B-MSC secretome without referring to exosomes and microvesicles (MVs) that can be secreted either on a constitutively basis or upon activation [37]. Exosomes are nanovesicles deriving from intracellular formations called multivesicular bodies (MVBs), which eventually fuse to the cellular membrane and release their exosome content in the extracellular space. They are found in various biological fluids, and it is believed that in this way they can **travel** from their secretion site and reach distant target cells. Microvesicles can contain all the molecules that are present in the parental cells (mRNA, microRNA (miRNA), lipids, proteins, etc.), and once taken up by the target cells, they transmit all kinds of cell-to-cell communication signals [38, 39, 40]. B-MSC MVs seem to have miRNAs that are not detectable in the parental cells while they seem to lack others that are, supporting the presence of a mechanism that controls the selective packing of cellular components into the MVs. When these selected sets of miRNAs are delivered to their target cells, they can suppress specific targets, most of which are related to the regulation of the immune system, development, cell survival, and cell differentiation. In addition to miRNAs, B-MSC MVs contain approximately700 proteins that can regulate self-renewal, proliferation, and tissue repair mechanisms [41].

In addition to exosomes and MVs, there have been a number of studies that report the transfer of mitochondria from B-MSCs. Spees and collaborators showed that human B-MSCs could rescue mitochondrial function in epithelial cells containing nonfunctional mitochondria [42]. This was confirmed recently in a mouse model of LPS-induced ARDS, where mitochondria from instilled human B-MSCs were visually tracked as they transferred to the mouse lung epithelium. The transfer seemed to be Cx43 dependent, and lung protection was abrogated when B-MSCs with defective mitochondria were used [43]. Whether the beneficial effects from B-MSCs are dependent on cell–cell contact to the lung epithelium or they are mainly achieved by the secretion of the paracrine factors is still not clear.

A feature that distinguishes B-MSCs from many cell types is that they can migrate to different tissues via blood circulation and home selectively to sites of injury [44, 45]. Presumably, the high concentration of inflammatory cytokines at the site of injury can attract B-MSCs, which are known to express a range of chemokine receptors (CCR2, CCR3, CCR4, and CCL5) and growth factor receptors (PDGF, IGF-1) [46]. Stromal cell-derived factor 1 (SDF-1) plays an important role in B-MSC homing through its interaction with CXCR4 receptor on the cell surface [47], while the secretion of metalloproteinases (MMP-2 and MT1-MMP) allows their extravasation and subendothelial migration [48].

In recent years, the emerging role of B-MSCs in ARDS models has unveiled more interesting functions of these cells. B-MSCs can **reprogram** macrophages into an anti-inflammatory and more phagocytic state, which would result in more efficient bacterial clearance [49, 50]. Moreover, they can directly inhibit bacterial growth by secreting antimicrobial molecules such as the peptide LL-37 [51] or the protein lipocalin 2 (LCN2) [50]. Through several studies, it has become obvious that in the case of ARDS, the beneficial effect of B-MSCs is mediated not only by modulating inflammation but also by protecting and restoring the local endothelium and epithelium integrity and permeability [45, 52, 53]. By combining cell and gene therapy, Xu et al. reported that the protective effect of B-MSCs in the LPS-injured mouse lung was greatly augmented by infusion of B-MSCs overexpressing angiopoietin 1 (ANGPT1), a vasculoprotective gene. B-MSC infusion significantly decreased airspace neutrophil and pulmonary vascular endothelial permeability after LPS administration [54]. As for epithelium integrity, keratinocyte growth factor (KGF) has been investigated as a mediating factor. KGF is secreted by B-MSCs as a soluble factor, and it participates in the regulation of the alveolar fluid transport. When KGF expression was inhibited on B-MSCs by iRNA treatment, the beneficial effects of B-MSCs or their conditional medium was diminished by 80% [52].

12.3 Impact of aging on mesenchymal stem cells

12.3.1 *In vitro* aging of MSCs

The low frequency of B-MSCs in bone marrow tissue (0.01% to 0.001%) necessitates there *in vitro* culture and expansion prior to any clinical application. Obviously, the conditions under which the cells are cultured have a great impact only on the lifespan but also the molecular profiles and the functional properties of these cells [55]. It is well established, for example, that low glucose concentration in the culture media is beneficial for the proliferative and differentiation potential of cultured B-MSCs [56]. Although it is generally understood that not all B-MSCs maintain their **stemness** in prolonged culture, only very recently, it was demonstrated that donor age has a critical impact on the kinetics and the stemlike properties of B-MSC cultures [57]. Long-term culture of B-MSCs from children, adults, and old individuals showed a decline in both the proliferation and differentiation of these cells with both passage number and donor age. B-MSCs from children and adult donors reached passage P24 and P15 before they stopped proliferating and maintained their spindle-like morphology until P15 and P9, respectively. On the contrary, B-MSCs from old donors reached only passage P7 before they stopped proliferating. Interestingly, even at a low passage when all the groups maintain their B-MSC morphology, the osteogenic and adipogenic potential of the cells is declining significantly with age, suggesting that there are underlying age-related differences on a subcellular level that could define the properties of these cells [58].

12.3.2   Age-related changes in B-MSCs

B-MSCs from elderly people are not only less in number, but they are bigger, they lose the characteristic spindle-like morphology of young B-MSCs [59], they form more podia, and they contain more actin stress fibers. Furthermore, they show a decline in their replicative capacity [59] and a slower proliferation rate in culture [60, 61], and a big proportion of them stain positive for senescence-associated beta-galactosidase [62]. B-MSCs from old mice are characterized by a quiescent state with low metabolic activity and are primarily in the G0 phase of the cell cycle. This quiescent state is maintained by both extrinsic and intrinsic mechanisms and has been postulated to be a way of preserving their long-term proliferative potential and genomic integrity [63, 64]. The problem that arises is that quiescent B-MSCs escape DNA damage checkpoints and several repair pathways that are cell cycle dependent and that have as a result the accumulation of DNA damage during aging, ultimately leading to rapid stem cell depletion or exhaustion. On a molecular level, aged B-MSCs show altered gene and miRNA expression profiles where key molecules of apoptosis, inflammation, self-renewal, proliferation, and differentiation are affected (see Table 12.1) [65, 66, 67]. In addition, aged B-MSCs show an increased production of reactive oxygen species (ROS) and oxidative damage [68], DNA-methylation changes affecting cell differentiation [69], and shorter telomeres [60].

12.3.3   Aging of B-MSCs versus aging of the organism

One of the biggest challenges in studying B-MSC biology and aging is the recreation of the bone marrow environment which consists of multiple cell types and matrix components that support a complex network of interactions with the B-MSCs. For practical purposes, most of these studies are performed using monolayer cultures of purified B-MSCs, which obviously lack the complexity of the *in vivo* system and make the investigation of the extrinsic factors that contribute to B-MSC aging quite unattainable. Nevertheless, several findings, which suggest that the age of the microenvironment of B-MSCs could be more important or as important as the age of the actual B-MSCs, support the theory of **extrinsic B-MSC aging**. Friedenstein et al. showed that the number of colony-forming units (CFU) from old transplants was increased by threefold when the transplants were previously implanted in young mice versus old mice [74]. Even in the case of the parabiotic pairing of old and young mice by Conboy that significantly improved muscle regeneration in the old mice, it was shown that pairing restored the expression of Delta-1 and Notch signaling in the skeletal muscle stem cells of the old animals, resulting in the restoration of their regenerative capacity which was compromised as animals aged [75, 76]. Parabiosis studies demonstrated that a broad improvement in tissue maintenance and repair could be promoted in an old mammal by factors from the young parabiont by boosting the regenerative capacity of tissue stem cells [77].

Table 12.1 Age-related molecular changes in B-MSCs.

Molecule	Organism	Change with age	Publication
IκB, interleukin-1α, iNOS, mitogen-activated protein kinase/ p38, ERK1/2, c-fos, c-jun	Human	Down	[66]
NF-κB, myc, interleukin-4 receptor	Human	Up	[66]
WNT3A, SFRF4, FZD8, CTNNB1, LEF1	Human	Down	[3]
FZD6, LRP6, FOSL1	Human	Up	[3]
HSP70, HSF1	Rhesus macaque	Down	[67]
TGFbeta1	Mouse	Down	[70]
p53, p21		Up	[68]
Trp53, Cdkn1, Rb1, Ccdne1, Ccdnd1, Mdm2, Chek2, Fas, Bax, Bad, Casp8, Apaf1, IGF-1, VEGF, HGF, Angpt1	Mouse	Down	[71]
E2F1, Brca1, Myc, Flt1	Mouse	Up	[71]
CCNE1, ERCC2, TGM2, RPL29	Human	Down	[72]
EPB41L3, TCEAL7, IL13RA2, MFAP5, ROBO1, S100A4, STEAP3, UBE2E2, CCND2, CDKN2B, DCN, PODN, TP53INP1, DRAM1, TGFBR3, RPS6KA2, PTGER4, FBXO32, SULF1, DBC1, HEXA, HEXB, ADAMTS5, COL8A2, GPNMB, TNFAIP6, CTSK, SPP1	Human	Up	[72]
BMP2/4	Mouse	Down	[70]
IL-6	Human	Up	[73]
RUNX2, Osterix, ALKP, BSP, OC	Human		[62]
mir-766, mir-558	Rhesus macaque	Up	[67]
mir-125b, mir-let-7f, mir-222, mir-221, mir-199-3p, mir-23a	Rhesus macaque	Down	[67]
45 miRNAs	Human	Differentially expressed	[66]

12.4 B-MSCs in disease

In recent years, several studies have shown that the number and the functions of adult stem cells deteriorate with age, but in most cases, it remains unclear whether these changes are the cause of aging or they are just a consequence of aging that should be viewed simply as a marker.

For example, old mice show an increased sensitivity to bleomycin injury and gammaherpesvirus infection and a higher susceptibility to LPS-induced acute lung injury (Rojas unpublished data). Is the severity of the disease symptoms augmented by age-compromised endogenous B-MSCs, or is the inflicted damage more severe and somehow less manageable in the older animals? To investigate the relationship between aging B-MSCs and severity of symptoms, we designed experiments (Rojas unpublished data), where LPS (1 mg/kg) was administered to old (2 years old) and young (3 months old) mice to induce ARDS; the animals were sacrificed at 48 h to collect blood, BAL, and lung samples for analysis. We observed that young mice had less inflammation, lower macrophages/lymphocytes/neutrophils counts, and less edema, suggesting an increase in susceptibility to ARDS associated with age. Although probably not the sole reason, we hypothesized that endogenous B-MSCs could be functionally compromised because of their age and thus contribute to the more severe disease phenotype. To test whether

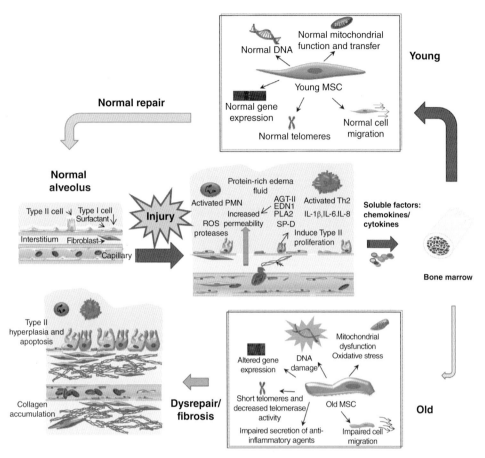

Figure 12.1 Comparison of the mechanisms of response in young and aged individuals. In the normal alveolus (left), there is normal fluid movement from the vascular to the interstitial space, with normal architecture of the epithelium and the endothelium, and production of surfactant. The panel in the middle shows an injured alveolus. Injury activates macrophages, which secrete a battery of proinflammatory cytokines and chemokines that leak into the bloodstream. Among these cytokines, IL-8 is responsible for the recruitment of neutrophils into the injured alveolus. ROS and proteases released by the activated neutrophils damage the alveolar epithelium and endothelium, and as a result, there is an increase in the permeability and in the influx of protein-rich edema. Another characteristic during injury is the increased levels of surfactant protein D by type II alveolar epithelial cells. The top right panel shows the representation of a restored alveolus including recruitment of healthy B-MSC. MSCs themselves produce and stimulate the production of IL-10 by macrophages, which attenuates their inflammatory cascade. The secretion of angiopoietin 1 and the transfer of mitochondria to the epithelial cells by B-MSC together with the production of KGF contribute to restoring the epithelium and endothelium. This results in improved alveolar fluid clearance and decreased permeability. The bottom right panel shows a defective senescent B-MSC with a decrease in their ability to respond to activation and mobilization, resulting in lung disrepair and fibrosis. To see a color version of this figure, see Plate 12.1.

old B-MSCs can modulate the severity of ARDS symptoms, we administered B-MSCs from old (2 years old) and young (3 months old) mice to LPS-treated mice (3 months old). Our results show that mice infused with young B-MSCs have lower levels of inflammatory cytokines in their plasma, lower macrophages/lymphocytes/neutrophils counts, and less edema, suggesting a causative link between aged B-MSCs and persistent lung inflammation and disrepair (Figure 12.1).

There are multiple studies that provide ample evidence regarding the cellular pathways that are crucial for proper stem cell function, such as DNA repair and telomere maintenance, but unfortunately without questioning the impact that a dysfunction in any of these B-MSC pathways would have on aging. A recent study provides the first direct evidence that the accumulation of DNA damage and loss of adult stem cell function in aging B-MSCs contributes to aging. In this study, the administration of muscle-derived stem/progenitor cells (MDSPCs) from young wild-type mice but not from old and progeroid (Ercc1−/− and Ercc1−/Δ) mice results in significant improvement of lifespan and health of progeroid mice. Moreover, the same study explores the mechanism by which MDSPCs confer their beneficial effects and argues toward a paracrine mechanism that supports tissue regeneration [43]. Ercc1 protein has a function in nucleotide excision repair of damaged DNA. DNA repair capacity is reduced with aging, and an inverse association between age and Ercc1 expression has been found in peripheral blood cells. Patients older than 50 years old have lower Ercc1 expression. In a parallel study (unpublished data), we have used the same progeria model to demonstrate that Ercc1−/Δ 12-week-old mice have an increased susceptibility to injury and lung fibrosis after bleomycin treatment. Ercc1−/Δ mice developed more severe pneumonitis and collagen deposition than controls and an increase in the number of cells in the BAL with elevated expression of IL-6 and TNFα in plasma. A direct link between the defective Ercc1-related B-MSCs and the lifespan of the Ercc1−/Δ mice was established using a parabiosis system pairing a young WT with an Ercc1−/Δ mouse, which resulted in an increased lifespan and body weight of the progeric animals. Furthermore, infusion of young WT B-MSCs but not natural old B-MSCs or murine embryonic fibroblasts increases the lifespan of the more severe Ercc1 mutants (Ercc1−/−).

In a rat model for cardiomyopathy, human B-MSC from aged donors showed a significant decrease of cardiac regenerative capacity. Briefly, injection of human B-MSCs from young (1–5 years old) but not from old donors (50–70 years old) into rat infarcted myocardia resulted in a substantial restoration of cardiac function along with an increase in vascular density and a decrease in matrix degradation [78]. In another case, the administration of stem cells from young mice (2 months old) but not from old mice (18 months old) improved cardiac function and restored cardiac angiogenesis in senescent mice after myocardial infarction [79]. B-MSCs from old donors (>45 years old) fail to differentiate *in vitro* into neuroectodermal cells [80], and early passage B-MSCs are more efficient in promoting the proliferation and maintenance of hematopoietic progenitor cells when cocultured *in vitro* [81].

12.5   B-MSCs in therapy

Several of B-MSCs' qualities make them very attractive therapeutic choices for a vast range of diseases. Despite their appeal, it is important to remember that multiple issues have to be addressed and evaluated in order to ensure their efficiency and safety in clinical applications.

12.5.1   *Ex vivo* expansion

As already mentioned, the low frequency of B-MSCs in bone marrow aspirates necessitates the *ex vivo* expansion of the carefully isolated B-MSCs. Several methods have been described for efficiently expanding B-MSCs, but what is important to remember is that the ultimate goal is to acquire a population of B-MSCs that are as close to the initial native state as possible.

A significant problem is that as the cells advance in passage number promoting cell senescence with the ultimate consequence that B-MSCs undergo both morphological and functional changes that reduce their stemlike behavior and consequently their clinical potential [82, 83].

12.5.2 Conditions affecting the expansion

(i) Advanced age of the donor has been reported to adversely affect lifespan/senescence, PD rates, and differentiation potential of cultured B-MSCs; (ii) donor's alcohol consumption produces an aged-like phenotype which predisposes cultured B-MSCs toward differentiation into the adipogenic lineage; and (iii) preexisting medical conditions such as diabetes, obesity, osteoporosis, and amyotrophic lateral sclerosis also compromise the differentiation potential of B-MSC cultures.

12.5.3 Autologous versus allogeneic B-MSCs

It is generally accepted, albeit there is some controversy in the past years, that B-MSCs can escape allogeneic rejection, and most of the clinical studies have shown that the infusion of allogeneic B-MSCs does not instigate any adverse reaction from the host [83].

12.5.4 Combination of cell preparations

Although the majority of the clinical approaches focus on the delivery of a pure B-MSC preparation, there is an emerging set of clinical studies that view the stem cell as a combination system that can be augmented by the codelivery of another cell type or a growth factor.

12.5.5 Delivery and targeting

The delivery method should ensure that transfused B-MSCs reach the target tissue. Several options exist (intracoronary, intramyocardial, intrathecal, intratracheal, intraperitoneal), but intravenous infusion is the most common method. A very important parameter in targeting the infused B-MSCs to the right repair site is to ensure that B-MSCs retain the ability to migrate following the signals that are emitted from the damaged sites and efficiently home there. Our preliminary data suggest a decrease on the ability of aged B-MSCs to respond to injury as a consequence of a downregulation on the expression of several chemokine receptors. A better understanding of the mechanisms by which the B-MSCs are mobilized and homed and achieve their beneficial effects would allow manipulations toward the development of a more efficient and safe clinical approach.

12.6 Conclusion

B-MSCs are critical to promote and coordinate repair responses in the lung (as summarized in Figure 12.1). Aging and extensive *in vitro* culture of B-MSCs define telomeric length, pluripotency, proliferating potential, and overall their ability to execute their regenerative role [59, 84]. Thus, increased susceptibility to lung diseases with age could be the result of functional exhaustion and aging of B-MSCs. The use of B-MSCs as a therapeutic approach for several lung diseases is appealing given their immunomodulatory and repair capacity. The fact that B-MSCs from HLA-mismatched third-party donors have been shown to be equally effective to autologous cells opens the possibility that B-MSCs from younger donors can be used in patients of advanced age. Further studies are necessary to characterize completely the effect of aging in the number, recruitment capacity, and function of B-MSCs and their contributing role in the pathogenesis of common lung diseases.

Acknowledgments

The authors would like to thank Nayra Cardenes for her help with the design of the figure.

References

1. Kapetanaki MG, Mora AL, Rojas M. Influence of age on wound healing and fibrosis. *J Pathol* 2013;229:310–322.
2. Sueblinvong V, Neujahr DC, Mills ST, et al. Predisposition for disrepair in the aged lung. *Am J Med Sci* 2012;344:41–51.
3. Faner R, Rojas M, Macnee W, Agusti A. Abnormal lung aging in chronic obstructive pulmonary disease and idiopathic pulmonary fibrosis. *Am J Respir Crit Care Med* 2012;186:306–313.
4. Leslie KO. Idiopathic pulmonary fibrosis may be a disease of recurrent, tractional injury to the periphery of the aging lung: A unifying hypothesis regarding etiology and pathogenesis. *Arch Pathol Lab Med* 2012;136:591–600.
5. Schraufnagel DE. Breathing in America: Diseases, progress, and hope. New York: American Thoracic Society; 2010.
6. Lozano R, Naghavi M, Foreman K, et al. Global and regional mortality from 235 causes of death for 20 age groups in 1990 and 2010: A systematic analysis for the global burden of disease study 2010. *Lancet* 2012;380:2095–2128.
7. Roberts MH, Dalal AA. Clinical and economic outcomes in an observational study of COPD maintenance therapies: Multivariable regression versus propensity score matching. *Int J Chron Obstruct Pulmon Dis* 2012;7:221–233.
8. Fernandez Perez ER, Daniels CE, Schroeder DR, et al. Incidence, prevalence, and clinical course of idiopathic pulmonary fibrosis: A population-based study. *Chest* 2010;137:129–137.
9. Blank R, Napolitano LM. Epidemiology of ARDS and ALI. *Crit Care Clin* 2011;27:439–458.
10. Dushianthan A, Grocott MP, Postle AD, Cusack R. Acute respiratory distress syndrome and acute lung injury. *Postgrad Med J* 2011;87:612–622.
11. Bhadade RR, de Souza RA, Harde MJ, Khot A. Clinical characteristics and outcomes of patients with acute lung injury and ards. *J Postgrad Med* 2011;57:286–290.
12. Vincent JL, Sakr Y, Groeneveld J, et al. ARDS of early or late onset: Does it make a difference? *Chest* 2010;137:81–87.
13. Alder JK, Chen JJ, Lancaster L, et al. Short telomeres are a risk factor for idiopathic pulmonary fibrosis. *Proc Natl Acad Sci USA* 2008;105:13051–13056.
14. Alder JK, Guo N, Kembou F, et al. Telomere length is a determinant of emphysema susceptibility. *Am J Respir Crit Care Med* 2011;184:904–912.
15. Amsellem V, Gary-Bobo G, Marcos E, et al. Telomere dysfunction causes sustained inflammation in chronic obstructive pulmonary disease. *Am J Respir Crit Care Med* 2011;184:1358–1366.
16. Aoshiba K, Nagai A. Oxidative stress, cell death, and other damage to alveolar epithelial cells induced by cigarette smoke. *Tob Induc Dis* 2003;1:219–226.
17. Xu J, Gonzalez ET, Iyer SS, et al. Use of senescence-accelerated mouse model in bleomycin-induced lung injury suggests that bone marrow-derived cells can alter the outcome of lung injury in aged mice. *J Gerontol* 2009;64:731–739.
18. Torres-Gonzalez E, Bueno M, Tanaka A, et al. Role of endoplasmic reticulum stress in age-related susceptibility to lung fibrosis. *Am J Respir Cell Mol Biol* 2012;46:748–756.

19. Wang JL, Wang PC. The effect of aging on the DNA damage and repair capacity in 2BS cells undergoing oxidative stress. *Mol Biol Rep* 2012;39:233–241.

20. Vidal MA, Walker NJ, Napoli E, Borjesson DL. Evaluation of senescence in mesenchymal stem cells isolated from equine bone marrow, adipose tissue, and umbilical cord tissue. *Stem Cells Dev* 2012;21:273–283.

21. Pittenger MF, Mackay AM, Beck SC, et al. Multilineage potential of adult human mesenchymal stem cells. *Science* 1999;284:143–147.

22. Weissman IL. Stem cells: Units of development, units of regeneration, and units in evolution. *Cell* 2000;100:157–168.

23. Singer NG, Caplan AI. Mesenchymal stem cells: Mechanisms of inflammation. *Annu Rev Pathol* 2011;6:457–478.

24. Dominici M, Le Blanc K, Mueller I, et al. Minimal criteria for defining multipotent mesenchymal stromal cells. The international society for cellular therapy position statement. *Cytotherapy* 2006;8:315–317.

25. Uccelli A, Moretta L, Pistoia V. Mesenchymal stem cells in health and disease. *Nat Rev Immunol* 2008;8:726–736.

26. Cardenes N, Caceres E, Romagnoli M, Rojas M. Mesenchymal stem cells: A promising therapy for the acute respiratory distress syndrome. *Respiration* 2013;85:267–278.

27. Lee JW, Zhu Y, Matthay MA. Cell-based therapy for acute lung injury: Are we there yet? *Anesthesiology* 2012;116:1189–1191.

28. Alessandra C, Andrea B, Floriana M, et al. Polyamines modulate epithelial-to-mesenchymal transition. *Amino Acids* 2012;42:783–789.

29. Aggarwal S, Pittenger MF. Human mesenchymal stem cells modulate allogeneic immune cell responses. *Blood* 2005;105:1815–1822.

30. Kawashima M, Kawakita T, Higa K, et al. Subepithelial corneal fibrosis partially due to epithelial-mesenchymal transition of ocular surface epithelium. *Mol Vis* 2010;16:2727–2732.

31. Kibria R, Sharma K, Ali SA, Rao P. Upper gastrointestinal bleeding revealing the stomach metastases of renal cell carcinoma. *J Gastrointest Cancer* 2009;40:51–54.

32. Matysiak M, Stasiolek M, Orlowski W, et al. Stem cells ameliorate EAE via an indoleamine 2,3-dioxygenase (IDO) mechanism. *J Neuroimmunol* 2008;193:12–23.

33. Nasef A, Chapel A, Mazurier C, et al. Identification of IL-10 and TGF-beta transcripts involved in the inhibition of T-lymphocyte proliferation during cell contact with human mesenchymal stem cells. *Gene Expr* 2007;13:217–226.

34. Song YS, Lee HJ, Doo SH, et al. Mesenchymal stem cells overexpressing hepatocyte growth factor (HGF) inhibit collagen deposit and improve bladder function in rat model of bladder outlet obstruction. *Cell Transplant* 2012;21:1641–1650.

35. Najar M, Raicevic G, Boufker HI, et al. Mesenchymal stromal cells use PGE2 to modulate activation and proliferation of lymphocyte subsets: Combined comparison of adipose tissue, Wharton's Jelly and bone marrow sources. *Cell Immunol* 2010;264:171–179.

36. de Pablo R, Monserrat J, Reyes E, et al. Sepsis-induced acute respiratory distress syndrome with fatal outcome is associated to increased serum transforming growth factor beta-1 levels. *Eur J Intern Med* 2012;23:358–362.

37. Baglio SR, Pegtel DM, Baldini N. Mesenchymal stem cell secreted vesicles provide novel opportunities in (stem) cell-free therapy. *Front Physiol* 2012;3:359.

38. Kangelaris KN, Sapru A, Calfee CS, et al. The association between a Darc gene polymorphism and clinical outcomes in African American patients with acute lung injury. *Chest* 2012;141:1160–1169.

39. Mommsen P, Zeckey C, Andruszkow H, et al. Comparison of different thoracic trauma scoring systems in regards to prediction of post-traumatic complications and outcome in blunt chest trauma. *J Surg Res* 2012;176:239–247.

40. Kinikar AA, Kulkarni RK, Valvi CT, et al. Predictors of mortality in hospitalized children with pandemic H1N1 influenza 2009 in Pune, India. *Indian J Pediatr* 2012;79:459–466.

41. Kim HS, Choi DY, Yun SJ, et al. Proteomic analysis of microvesicles derived from human mesenchymal stem cells. *J Proteome Res* 2012;11:839–849.

42. Spees JL, Olson SD, Whitney MJ, Prockop DJ. Mitochondrial transfer between cells can rescue aerobic respiration. *Proc Natl Acad Sci USA* 2006;103:1283–1288.

43. Lavasani M, Robinson AR, Lu A, et al. Muscle-derived stem/progenitor cell dysfunction limits healthspan and lifespan in a murine progeria model. *Nat Commun* 2012;3:608.

44. Ortiz LA, Gambelli F, McBride C, et al. Mesenchymal stem cell engraftment in lung is enhanced in response to bleomycin exposure and ameliorates its fibrotic effects. *Proc Natl Acad Sci USA* 2003;100:8407–8411.

45. Rojas M, Xu J, Woods CR, et al. Bone marrow-derived mesenchymal stem cells in repair of the injured lung. *Am J Respir Cell Mol Biol* 2005;33:145–152.

46. Ponte AL, Marais E, Gallay N, et al. The in vitro migration capacity of human bone marrow mesenchymal stem cells: Comparison of chemokine and growth factor chemotactic activities. *Stem Cells* 2007;25:1737–1745.

47. Wynn RF, Hart CA, Corradi-Perini C, et al. A small proportion of mesenchymal stem cells strongly expresses functionally active CXCR4 receptor capable of promoting migration to bone marrow. *Blood* 2004;104:2643–2645.

48. Ries C, Egea V, Karow M, Kolb H, Jochum M, Neth P. MMP-2, MT1-MMP, and TIMP-2 are essential for the invasive capacity of human mesenchymal stem cells: Differential regulation by inflammatory cytokines. *Blood* 2007;109:4055–4063.

49. Lee JW, Krasnodembskaya A, McKenna DH, Song Y, Abbott J, Matthay MA. Therapeutic effects of human mesenchymal stem cells in ex vivo human lungs injured with live bacteria. *Am J Respir Crit Care Med* 2013;187:751–760.

50. Conrad SA, Rycus PT, Dalton H. Extracorporeal life support registry report 2004. *ASAIO J* 2005;51:4–10.

51. Wei X, Dombkowski D, Meirelles K, et al. Mullerian inhibiting substance preferentially inhibits stem/progenitors in human ovarian cancer cell lines compared with chemotherapeutics. *Proc Natl Acad Sci USA* 2010;107:18874–18879.

52. Lee JW, Fang X, Gupta N, Serikov V, Matthay MA. Allogeneic human mesenchymal stem cells for treatment of E. coli endotoxin-induced acute lung injury in the ex vivo perfused human lung. *Proc Natl Acad Sci USA* 2009;106:16357–16362.

53. Gupta N, Su X, Popov B, Lee JW, Serikov V, Matthay MA. Intrapulmonary delivery of bone marrow-derived mesenchymal stem cells improves survival and attenuates endotoxin-induced acute lung injury in mice. *J Immunol* 2007;179:1855–1863.

54. Xu J, Qu J, Cao L, et al. Mesenchymal stem cell-based angiopoietin-1 gene therapy for acute lung injury induced by lipopolysaccharide in mice. *J Pathol* 2008;214:472–481.

55. Roobrouck VD, Vanuytsel K, Verfaillie CM. Concise review: Culture mediated changes in fate and/or potency of stem cells. *Stem Cells* 2011;29:583–589.

56. Stolzing A, Scutt A. Age-related impairment of mesenchymal progenitor cell function. *Aging Cell* 2006;5:213–224.

57. Zaim M, Karaman S, Cetin G, Isik S. Donor age and long-term culture affect differentiation and proliferation of human bone marrow mesenchymal stem cells. *Ann Hematol* 2012;91:1175–1186.

58. Kim M, Kim C, Choi YS, Park C, Suh Y. Age-related alterations in mesenchymal stem cells related to shift in differentiation from osteogenic to adipogenic potential: Implication to age-associated bone diseases and defects. *Mech Ageing Dev* 2012;133:215–225.

59. Baxter MA, Wynn RF, Jowitt SN, Wraith JE, Fairbairn LJ, Bellantuono I. Study of telomere length reveals rapid aging of human marrow stromal cells following in vitro expansion. *Stem Cells* 2004;22:675–682.

60. Choumerianou DM, Martimianaki G, Stiakaki E, Kalmanti L, Kalmanti M, Dimitriou H. Comparative study of stemness characteristics of mesenchymal cells from bone marrow of children and adults. *Cytotherapy* 2010;12:881–887.

61. Roobrouck VD, Ulloa-Montoya F, Verfaillie CM. Self-renewal and differentiation capacity of young and aged stem cells. *Exp Cell Res* 2008;314:1937–1944.

62. Zhou S, Greenberger JS, Epperly MW, et al. Age-related intrinsic changes in human bone-marrow-derived mesenchymal stem cells and their differentiation to osteoblasts. *Aging Cell* 2008;7:335–343.

63. Washko GR, Hunninghake GM, Fernandez IE, et al. Lung volumes and emphysema in smokers with interstitial lung abnormalities. *N Engl J Med* 2011;364:897–906.

64. Burgel PR. [pathogenesis of chronic obstructive pulmonary disease]. *Presse Med* 2009;38:406–412.

65. Guo L, Zhao RC, Wu Y. The role of microRNAs in self-renewal and differentiation of mesenchymal stem cells. *Exp Hematol* 2011;39:608–616.

66. Pandey AC, Semon JA, Kaushal D, et al. MicroRNA profiling reveals age-dependent differential expression of nuclear factor kappaB and mitogen-activated protein kinase in adipose and bone marrow-derived human mesenchymal stem cells. *Stem Cell Res Ther* 2011;2:49.

67. Yu JM, Wu X, Gimble JM, Guan X, Freitas MA, Bunnell BA. Age-related changes in mesenchymal stem cells derived from rhesus macaque bone marrow. *Aging Cell* 2011;10:66–79.

68. Stolzing A, Jones E, McGonagle D, Scutt A. Age-related changes in human bone marrow-derived mesenchymal stem cells: Consequences for cell therapies. *Mech Ageing Dev* 2008;129:163–173.

69. Bork S, Pfister S, Witt H, et al. DNA methylation pattern changes upon long-term culture and aging of human mesenchymal stromal cells. *Aging Cell* 2010;9:54–63.

70. Umberto Meduri G, Bell W, Sinclair S, Annane D. Pathophysiology of acute respiratory distress syndrome. Glucocorticoid receptor-mediated regulation of inflammation and response to prolonged glucocorticoid treatment. *Presse Med* 2011;40:e543–e560.

71. Wilson A, Shehadeh LA, Yu H, Webster KA. Age-related molecular genetic changes of murine bone marrow mesenchymal stem cells. *BMC Genomics* 2010;11:229.

72. Jiang SS, Chen CH, Tseng KY, et al. Gene expression profiling suggests a pathological role of human bone marrow-derived mesenchymal stem cells in aging-related skeletal diseases. *Aging* 2011;3:672–684.

73. Cheleuitte D, Mizuno S, Glowacki J. In vitro secretion of cytokines by human bone marrow: Effects of age and estrogen status. *J Clin Endocrinol Metab* 1998;83:2043–2051.

74. Friedenstein AJ, Latzinik NV, Gorskaya YuF, Luria EA, Moskvina IL. Bone marrow stromal colony formation requires stimulation by haemopoietic cells. *Bone Miner* 1992;18:199–213.

75. Conboy IM, Rando TA. Aging, stem cells and tissue regeneration: Lessons from muscle. *Cell Cycle* 2005;4:407–410.

76. Galvin JR, Franks TJ. Smoking-related lung disease. *J Thorac Imaging* 2009;24:274–284.

77. Conboy IM, Rando TA. Heterochronic parabiosis for the study of the effects of aging on stem cells and their niches. *Cell Cycle* 2012;11:2260–2267.

78. Fan M, Chen W, Liu W, et al. The effect of age on the efficacy of human mesenchymal stem cell transplantation after a myocardial infarction. *Rejuvenation Res* 2010;13:429–438.

79. Khan M, Mohsin S, Khan SN, Riazuddin S. Repair of senescent myocardium by mesenchymal stem cells is dependent on the age of donor mice. *J Cell Mol Med* 2011;15:1515–1527.

80. Hermann A, List C, Habisch HJ, et al. Age-dependent neuroectodermal differentiation capacity of human mesenchymal stromal cells: Limitations for autologous cell replacement strategies. *Cytotherapy* 2010;12:17–30.

81. Walenda T, Bork S, Horn P, et al. Co-culture with mesenchymal stromal cells increases proliferation and maintenance of haematopoietic progenitor cells. *J Cell Mol Med* 2010;14:337–350.

82. Pal R, Hanwate M, Jan M, Totey S. Phenotypic and functional comparison of optimum culture conditions for upscaling of bone marrow-derived mesenchymal stem cells. *J Tissue Eng Regen Med* 2009;3:163–174.

83. Minguell JJ, Allers C, Lasala GP. Mesenchymal stem cells and the treatment of conditions and diseases: The less glittering side of a conspicuous stem cell for basic research. *Stem Cells Dev* 2013;22:193–203.

84. Guillot PV, Gotherstrom C, Chan J, Kurata H, Fisk NM. Human first-trimester fetal MSC express pluripotency markers and grow faster and have longer telomeres than adult MSC. *Stem Cells* 2007;25:646–654.

13 COPD as a Disease of Premature Aging

Laurent Boyer, Jorge Boczkowski, and Serge Adnot

INSERM U955 and Département de Physiologie-Explorations Fonctionnelles, Hôpital Henri Mondor, Université Paris Est, Paris, France

13.1 Introduction

Chronic obstructive pulmonary disease (COPD) is of increasing prevalence in most countries and is expected to become the third leading cause of death worldwide by 2020 [1]. COPD is a chronic disease characterized by irreversible airflow obstruction that can lead to severe chronic respiratory insufficiency. Cigarette smoke is the main etiologic factor for COPD, which develops in 20–25% of smokers. Histopathologically, inflammation and remodeling of distal bronchi and pulmonary alveoli are major determinants of the progressive respiratory function decline and fixed airflow obstruction in COPD. In addition to these abnormalities, pulmonary vessel remodeling followed by the development of pulmonary hypertension (PH) is a common complication of COPD that contributes to a poor prognosis [2, 3]. Although COPD has an alarming societal impact, its pathophysiology remains incompletely understood.

COPD is an age-related disease for which new data indicate associations with telomere shortening and exaggerated cell senescence [4, 5]. Because short telomeres are associated with increased susceptibility to replicative cell senescence, one current hypothesis includes cell senescence among the mechanisms underlying the pathological alterations at the origin of COPD. Senescent alveolar epithelial and endothelial cells were recently identified in lung specimens from patients with emphysema, and *in vitro* studies of human alveolar epithelial cells and fibroblasts showed induction of cell senescence by cigarette smoke and inhibition of this effect by antioxidants [6, 7, 8]. Furthermore, several studies documented telomere shortening in peripheral leukocytes from patients with COPD, as well as increased susceptibility to senescence of lung vascular cells, pulmonary artery smooth muscle cells (PA-SMCs), and fibroblasts obtained during lung surgery from smokers with COPD [4, 5, 7, 9, 10].

The association of COPD with the development of comorbidities is well established [11]. One current hypothesis is that premature aging in COPD may not only constitute a major propeller of COPD progression but also contribute to the development of systemic complications of adverse prognostic significance such as cardiovascular disorders, weight loss,

Molecular Aspects of Aging: Understanding Lung Aging, First Edition. Edited by Mauricio Rojas, Silke Meiners and Claude Jourdan Le Saux.

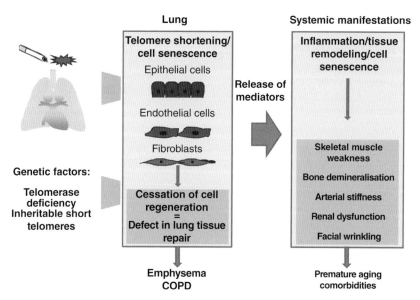

Figure 13.1 Diagram illustrating potential mechanisms linking cell senescence and development of lung alterations at the origin of COPD and systemic manifestations of the disease. To see a color version of this figure, see Plate 13.1.

bone demineralization, and muscle dysfunction. A unifying concept is that lung alterations contribute to drive the process of systemic senescence and premature aging in COPD, even in patients with only moderate degrees of lung dysfunction. Thus, elucidating the mechanisms that underlie premature aging in COPD would probably produce major clinical benefits.

Two concepts relevant to current research on these mechanisms are discussed in this chapter: (i) senescent cells contribute greatly to the pathogenesis of COPD, and (ii) lung alterations contribute to drive the process of systemic senescence and premature aging and the development of comorbidities in COPD (Figure 13.1).

13.2 Senescent cells contribute to the pathogenesis of COPD

13.2.1 Accumulation of senescent cells in COPD lungs

Senescent cells accumulate within tissues with advancing age. Somatic cell senescence occurs either when the replicative potential is exhausted or in response to excessive extracellular or intracellular stress [12]. Both forms of senescence may be accelerated in COPD: premature replicative senescence may result from increased telomere shortening and premature stress-related senescence from nontelomeric signals triggered by oxidative stress due primarily to cigarette smoke exposure. Telomeric signals are mediated chiefly via the p53–p21 pathway and nontelomeric signals via the p16–retinoblastoma protein pathway [13].

Increased numbers of p21- and p16-stained cells have been found in the lungs of patients with emphysema compared to control smokers [6, 14]. Previous reports of marked telomere shortening in patients with COPD are consistent with the increased number of senescent cells found in the lungs of patients with COPD compared to control smokers. Studies of human alveolar epithelial cells and fibroblasts also showed induction of cell senescence by cigarette smoke, as well as inhibition of this effect by antioxidants [7, 8].

Previous studies of lung specimens and cultured pulmonary vascular endothelial cells (P-ECs) derived from patients with COPD and from age- and sex-matched control smokers revealed that cultured P-ECs from patients with COPD displayed premature senescence compared to control smokers [9]. The number of senescent P-ECs in the lungs was higher in patients with COPD than in controls, and cultured P-ECs from patients with COPD displayed premature replicative senescence due to decreased telomerase activity with telomere short-ening. Suggestions about the mechanisms underlying premature P-EC senescence in COPD based on the results of these studies are still speculative. P-ECs from patients with COPD had decreased telomerase activity and reduced telomere length, together with increased p21 and p16 expression. These observations are consistent with a prominent role of increased cell turnover for the occurrence of replicative cell senescence in patients with COPD. In accordance with this possibility, an inverse relationship between the population doubling level (PDL) and telomere length was found at an early cell passage. However, p16 expression was also higher in cells from patients with COPD than in those from controls, suggesting a contribution of p16 in driving premature senescence in COPD. Thus, accelerated P-EC senescence in COPD may be attributable to a combination of both telomere shortening and oxidative stress responsible for p16 activation.

13.2.2 Inflammation and lung-cell senescence in COPD

Chronic inflammation is a prominent feature of COPD [15, 16, 17]. An exaggerated inflam-matory response of the airways to chronic irritants, primarily cigarette smoke, is considered the main cause of COPD. The inflammatory process results in airway remodeling and paren-chyma destruction. After smoking cessation, the inflammation persists, and the levels of proinflammatory cytokines in the lungs and bloodstream remain high in patients with COPD, even in those with mild and stable forms of the disease [18]. This persistent inflammation may not only constitute a major driver of COPD progression but also contribute to the development of the aforementioned systemic complications of adverse prognostic signifi-cance (cardiovascular disease, weight loss, bone demineralization, and muscle dysfunction) [19, 20]. The mechanisms of nonresolving inflammation in COPD remain incompletely understood. Elucidation of these mechanisms would probably lead to major improvements in the management of COPD.

Senescent cells that survive *in vivo* lose a number of functions and express several genes (encoding growth factors, proteases, and cytokines) involved in the remodeling of surround-ing tissues [21, 22]. Of note, these changes in gene expression resemble the chronic inflam-matory profile seen in COPD and are characterized by increased lung and bloodstream levels of various cytokines, proteases, and growth factors including IL6, CCL2, IL8, TGF-beta, and MMPs [21, 23, 24, 25]. Most of these mediators originate mainly from pulmonary vascular cells, that is, PA-SMCs and P-ECs. Previous studies also documented that cultured P-ECs undergoing replicative senescence released increased amounts of several cytokines and mediators and that this process was amplified in P-ECs from patients with COPD com-pared to controls. Various mediators were detected in P-EC-conditioned media from patients with COPD and controls: thus, IL6, IL8, RANTES, MCP-1, Hu-GRO, PDGF, bFGF, PAI-1, and sICAM-1 were detected using a Luminex® assay. Of note, most of these proteins secreted into P-EC-conditioned media from controls increased with the number of passages, indicating that their expression was linked to the normal process of replicative senescence [9]. Several of these factors, including IL6, IL8, Hu-GRO, MCP-1, and sICAM-1, were found in larger amounts at an early cell passage in P-EC-conditioned media from patients with COPD compared to controls, and most of them failed to increase further during subsequent cell passages [9]. Thus, the differences between patients with COPD and controls observed at an early cell passage but not at senescence were chiefly due to the larger proportion of senescent cells in patients with COPD. Accordingly, the amount of these mediators at an early P-EC passage correlated positively with the number of senescent β-galactosidase-positive cells and negatively with the PDL and telomere length [9].

Thus, premature senescence of P-ECs from patients with COPD was associated with marked overexpression of major proinflammatory cytokines and adhesion molecules, which affected monocyte adherence and migration [9]. Together with the elevated cytokine levels in the lungs from patients with COPD compared to controls and the correlation of these levels with telomere length and *in vitro* P-EC senescence criteria, these findings point to a direct link between lung inflammation in patients with COPD and premature P-EC senescence. The hypothesis that increased lung-cell senescence is sufficient to cause inflammation has received further support from studies in telomerase-deficient ($TR^{-/-}$) mice showing increased lung cytokine levels in proportion to the decrease in telomere length, even in the absence of external stimuli [9].

13.2.3 Emphysema and lung-cell senescence in COPD

According to a current hypothesis, emphysema lesions may result from an imbalance between the process of senescence or apoptosis of alveolar or endothelial cells, on the one hand, and the ability of these cells to proliferate and induce tissue repair, on the other. Thus, excessive senescence or apoptosis of alveolar or endothelial cells may cause the loss of alveolar structures, resulting in the development of emphysema lesions. According to this concept, young people may compensate for smoking-related lung-cell apoptosis or senescence by an increase in cell proliferation, thus preventing the development of emphysema lesions. In elderly individuals, cell senescence is accelerated, and the balance is tipped toward the formation of emphysema lesions [26]. That cell senescence may be the cause of insufficient cell proliferation seen in pulmonary emphysema is supported by several human and experimental studies [7, 27, 28, 29].

As mentioned earlier, severe lung emphysema in humans has been shown to be associated with an increased number of p21- and p16-stained alveolar epithelial and endothelial cells [6]. Interestingly, an inverse correlation was found in this study between p16 expression and proliferating cell nuclear antigen (PCNA) expression in alveolar epithelial cells and vascular endothelial cells, indicating that alveolar cell senescence is associated with decreases in cell proliferation and regeneration. In studies performed more recently in patients with COPD compared to control smokers, von Willebrand-positive P-ECs that stained for p21 and p16 were also detected, and cultured P-ECs derived from lung tissues showed characteristic features of accelerated senescence [9]. Thus, P-ECs from patients with COPD studied *in vitro* exhibited early replicative senescence compared to those from controls, with a marked decrease in the cumulative PDL and a higher percentage of β-galactosidase-positive cells at an early cell passage. In the overall cohort of patients with COPD and controls, a relationship was found between the PDL and the emphysema score, supporting a close association between cell senescence and COPD severity. However, given the mild degree of emphysema in this group of patients, it was difficult to establish a specific association between cell senescence and emphysema severity.

Telomere shortening has been suggested to promote the development of emphysema. In a family with idiopathic pulmonary fibrosis and a critical mutation in the telomerase gene, two family members who smoked cigarettes developed both idiopathic pulmonary fibrosis and early-onset emphysema [30]. Moreover, in a cohort of telomerase mutation carriers, 5% of cases were reported to have a history of spontaneous pneumothorax or to have COPD [31].

However, the strongest support for a cause–effect relationship between lung-cell senescence and emphysema comes from animal studies. In a study of two mouse strains, C57BL/6 with long telomeres and Cast/Eij with unusually short telomeres, each strain was crossed with $TR^{-/-}$ mice and then bred for four generations to induce further telomere shortening [30]. Of note, neither strain had abnormal lung structure or function in the absence of stress. However, in response to 6 months of low-level cigarette smoke exposure, $C57Bl/6xTR^{-/-}$ mice developed emphysema of greater severity compared to wild-type mice. Wild-type Cast/Eij mice, despite their relatively short telomeres, were resistant to cigarette smoke-induced emphysema, whereas $Cast/EijxTR^{-/-}$ mice developed significant airspace enlargement [30].

At the opposite, mechanisms protecting against lung-cell senescence may also protect against experimentally induced lung emphysema. An example is the anti-aging protein sirtuin-1 (SIRT1), which is overexpressed in the lungs of COPD patients. SIRT1 activation by genetic overexpression or by the selective pharmacological activator SIRT1720 attenuated stress-induced premature cell senescence and protected against emphysema induced by cigarette smoke and elastase in mice [28]. Interestingly, the effects of SIRT1 were shown to be mediated by FOXO3 in this study. Similarly, mice lacking p21 and exhibiting resistance to cell senescence were protected against cigarette smoke-induced emphysema [28]. A role for the mTOR pathways has also been suggested as a mechanism protecting against experimental cigarette smoke-induced lung emphysema. Indeed, transgenic mice exhibiting mTOR activation in alveolar epithelial cells develop lesser degree of cigarette smoke-induced emphysema than wild-type mice [32].

Most of these studies focused on the role of senescent epithelial cells as major mediators of emphysema development. For example, SIRT1 ablation in the airway epithelium but not in myeloid cells aggravated the airspace enlargement in mice, suggesting that emphysema in these models occurred independently from inflammation [28].

13.2.4 Pulmonary hypertension and cell senescence in COPD

COPD is one of the most common causes of PH and cor pulmonale [33]. Although PH is usually mild to moderate in COPD, its severity varies greatly across patients and its presence is associated with a poor prognosis. Extensive pulmonary vessel remodeling with prominent intimal thickening, medial hypertrophy, and small-arteriole muscularization are cardinal pathological features of PH in COPD. Although these structural changes are considered the major cause of the increase in pulmonary vascular resistance, their pathogenesis remains uncertain.

In a population of patients with COPD investigated by right heart catheterization, the extent of telomere shortening was associated with PH severity independently from the severity of airflow obstruction, age, and smoking history [10]. To evaluate whether pulmonary vessels from patients with COPD exhibited cell senescence and whether cell senescence was related to pulmonary vascular remodeling, pulmonary vessels and cultured PA-SMCs derived from patients with COPD and from sex- and age-matched control smokers were investigated [10]. Pulmonary vessels from patients with COPD exhibited increased wall hypertrophy compared to controls, in keeping with the higher pulmonary artery pressure in the patients with COPD than in the controls. Immunohistochemical examination of pulmonary vessels revealed an increased percentage of pulmonary vascular cells that stained positive for p21 and p16, including P-ECs and PA-SMCs, in patients with COPD compared to controls. These cells were identified as senescent cells by experiments performed in the proximal pulmonary arteries, where β-galactosidase-stained cells were also positive for p16 and p21. PA-SMCs derived from pulmonary vessels also exhibited characteristic features of accelerated senescence when studied *in vitro*. Cultured PA-SMCs from patients with COPD showed premature senescence when compared to those of controls, with a marked decrease in the cumulative PDL and a higher percentage of β-galactosidase-positive cells at an early passage.

Interestingly, remodeled vessels from patients with COPD contained more proliferating Ki67-stained PA-SMCs and accumulated extracellular matrix than did those from controls. Thus, remodeled vessels from patients with COPD were paradoxically characterized by a combination of elevated senescent-cell counts with an increased proportion of proliferating cells and increased extracellular matrix deposition. Of note, studies of remodeled vessels at sites of vascular hypertrophy revealed senescent cells to be virtually confined to the media, with only a few senescent cells in the neointima, whereas proliferating cells predominated in the neointima and hypertrophied media. These results support the concept that several PA-SMC subsets are present in the pulmonary vascular wall of remodeled vessels in COPD and that these subsets work in combination to participate in the remodeling process [14, 34]. Similar results have been reported in atherosclerotic lesions characterized by senescent cells and actively dividing cells in the neointima, possibly of monoclonal origin [35].

Identifying the exact mechanisms by which PA-SMC senescence contributes to pulmonary vascular remodeling is a challenging task. Proliferation of neighboring cells may occur via a direct mitogenic effect or via indirect effects mediated by tissue damage or inflammatory cell recruitment. As also found for other cell types, PA-SMCs undergoing replicative senescence released excessive amounts of several cytokines and mediators. The amount of these secreted factors differed markedly between patients with COPD and controls at an early cell passage but not at senescence, indicating that the differences were mainly due to the higher proportion of senescent cells in patients with COPD. Several factors such as IL6, IL8, MCP-1, and IL1β have been demonstrated to make a strong but indirect contribution to pulmonary vascular remodeling [36]. In particular, IL6 is a major contributor to hypoxic PH and is closely linked to PH severity in patients with COPD or idiopathic PH [37, 38].

Soluble and insoluble factors released by senescent cells were found to stimulate the growth and migration of target PA-SMCs. A similar finding was obtained previously using senescent fibroblasts and cultured epithelial cells and was taken as evidence that senescent cells promoted cell proliferation and tumor growth [39]. Thus, actively dividing cells in the neointimal lesions of remodeled vessels were found to be surrounded by senescent cells, suggesting cross talk between the two cell subsets [10]. In this study, cell proliferation in response to culture media from senescent cells was markedly reduced in the presence of neutralizing antibodies to IL6 and MCP-1; and cell migration was reduced in the presence of neutralizing antibodies to IL8, TNF-α, IL6, and MCP-1. The fact that excess PA-SMC proliferation in response to culture media from senescent compared to presenescent cells was no longer observed in the presence of anti-MCP-1 or anti-IL6 antibodies suggests an important role for these cytokines in this process. Moreover, these results suggest that the global action of these soluble factors may be stimulation of growth and migration and not stimulation of senescence. Thus, senescent cells may create a microenvironment that facilitates the migration and growth of nonsenescent cells, thereby inducing neointima formation and vessel remodeling. Whether the phenotype of nonsenescent cells involved in neointima formation is normal or abnormal remains to be elucidated.

13.3 Lung dysfunction and the general process of premature aging in COPD

COPD is a lung disease associated with systemic consequences and comorbidities. Patients with COPD are at risk for developing systemic alterations such as bone demineralization, muscle atrophy and weakness, weight loss, cardiovascular disease, kidney dysfunction, and insulin resistance. The clinical relevance of these systemic effects is now well established, as reflected by international guidelines for COPD diagnosis and treatment. Whether these alterations are related to a common pathogenic mechanism or develop independently in patients with COPD remains an open question.

One current concept is that systemic COPD complications reflect a global process of accelerated aging that develops as a consequence of the lung disease. Patients with COPD exhibit telomere shortening in circulating leukocytes compared to smokers without COPD, indicating that the accelerated aging process is not restricted to the lung but affects the entire body [4, 5]. Interestingly, there is a close relationship between cell senescence and inflammation, and there is now sound evidence that some of the inflammatory cytokines released by senescent cells can induce cell senescence, thereby propagating the aging process. Thus, the cell senescence and inflammatory processes are closely linked in COPD and may participate in the systemic manifestations of the disease and comorbidities (Figure 13.1).

13.3.1 Clinical manifestations of premature aging in COPD patients

The hypothesis that COPD is linked to premature aging is based on several studies showing that patients with COPD, compared to age- and sex-matched controls with similar smoking histories, display the aforementioned major systemic alterations (progressive bone demineralization, muscle atrophy and weakness, weight loss, cardiovascular alterations, kidney dysfunction, and insulin resistance) [11].

Among these alterations, muscle wasting appears as a major prognostic factor. In severe COPD, loss of body and muscle weight (pulmonary cachexia) significantly contributes to skeletal muscle weakness, impaired exercise capacity, decreased health status, and increased mortality, independently from primary organ failure [40]. Muscle wasting is associated with increased risks for hospital readmission after a COPD exacerbation and mechanical ventilation. Furthermore, muscle wasting has been identified as a significant determinant of mortality in COPD, independent from lung function, smoking, and body mass index [41, 42]. Bone demineralization is also a well-documented systemic manifestation of the disease, and osteoporosis is highly prevalent in patients with COPD, irrespective of gender [19]. In one study of COPD patients, 18% of men and 30% of women developed osteoporosis, and 42% of men and 41% of women had osteopenia [43]. The increase of arterial stiffness in patients with COPD compared to controls as assessed by pulse-wave velocity [19, 44] also suggests acceleration of the vascular aging process in COPD patients. This arterial stiffness increase is probably linked to the excess cardiovascular risk observed in COPD patients, particularly as arterial stiffness predicts cardiovascular events in the general population [45]. Recent work also suggests that the arterial stiffness associated with COPD is associated with increased calcification of large vessels, another feature of pathological aging [46, 47]. Other manifestations of aging including kidney dysfunction and insulin resistance occur in COPD patients [48]. The level of pulmonary emphysema has been shown to be associated with decreased glomerular filtration rates in cigarette smokers, independently from common risk factors for kidney disease; and an increased prevalence of diabetes has been reported among COPD patients (relative risk, 1.5–1.8), even those with mild disease [49]. Another interesting feature related to aging is the development of facial wrinkling that increases with the number of cigarette packs smoked per year [50]. Severe facial wrinkling is associated with COPD and correlates inversely with FEV1, with 80% of smokers who have facial wrinkling fulfilling spirometry criteria for COPD [51].

13.3.2 Role for lung alterations in systemic premature aging during COPD

COPD causes a range of lung manifestations, with variable combinations of parenchymal emphysema, small airway narrowing, and fibrosis. Several reports suggest that emphysema may be the strongest predictor of premature aging, as opposed to the severity of the obstruction evaluated based on the FEV1 decrease [48, 52]. Greater severity of emphysema is associated with renal dysfunction, osteoporosis/osteopenia, arterial stiffness, and facial wrinkling in COPD patients or smokers without COPD [48, 51, 52, 53]. Moreover, some studies suggest that emphysema alone may be an independent marker for several systemic manifestations of aging in patients with COPD. McAllister and colleagues demonstrated a strong, independent relationship between the extent of emphysema on quantitative CT analysis and arterial stiffness measured by pulse-wave velocity [53]. Likewise, Bon et al. documented an independent relationship between the presence of emphysema and bone mineral density or kidney function, which were not correlated with either FEV1 or quantitative CT scan measures of airway size [52]. Thus, emphysema appears to be an independent risk factor for many systemic manifestations of COPD, suggesting that common pathogenic mechanisms may lead to both parenchymal destruction and systemic aging.

Interestingly, Maclay et al. postulated that COPD might be associated with both local lung alterations and systemic connective-tissue abnormalities. They showed that patients with COPD had increased skin elastin degradation compared with control smokers, in relation to emphysema severity and pulse-wave velocity [54]. They proposed that elastin degradation – a feature of normal aging – occurring in elastin-rich structures such as the alveolar wall, walls of large conduit arteries, and skin might constitute a common denominator of the aging process in patients with emphysema. Various matrix proteinases are released by senescent cells and may account for these alterations, although they cannot explain bone demineralization or muscle wasting. Moreover, whether emphysema may trigger these processes or is one component of a more general aging process involving the lung alveoli remains unclear.

13.4 Conclusion

Conceptually, diminishing the accelerated cell senescence seen in COPD may improve both the lung alterations and the systemic age-related abnormalities and comorbidities. At present, efforts to reduce the chronic inflammation in patients with COPD have produced disappointing results. Experimentally, several therapeutic strategies have been successful in preventing or reversing the aging process. For instance, telomerase activators prolonged the life span, increased telomere length, and improved bone mineral density and insulin resistance in aged mice [55]. Also, statins combined with aminobisphosphonates have been shown to extend the life span and to improve the lipodystrophy, hair loss, and bone defects in a mouse model of human premature aging [56]. A pharmacological SIRT1 activator was recently shown to prevent cigarette smoke-induced emphysema in mice, whereas the inhibition of lung inflammation had no beneficial effect [28]. SIRT1 regulates the stress response transcription factor FOXO3 and may therefore modulate various aging-related alterations including those affecting the skeletal muscle function, cardiovascular homeostasis, and life span. Current therapies for COPD are symptomatic and have no effect on lung-cell maintenance. Whether new therapeutic strategies directed against the aging process may be of clinical benefit in patients with COPD remains unknown.

References

1. Celli BR, MacNee W. Standards for the diagnosis and treatment of patients with COPD: A summary of the ATS/ERS position paper. Eur Respir J 2004;23:932–946.
2. Minai OA, Chaouat A, Adnot S. Pulmonary hypertension in COPD: Epidemiology, significance, and management: Pulmonary vascular disease: The global perspective. Chest 2010;137:39S–51S.
3. Santos S, Peinado VI, Ramirez J, et al. Characterization of pulmonary vascular remodelling in smokers and patients with mild COPD. Eur Respir J 2002;19:632–638.
4. Savale L, Chaouat A, Bastuji-Garin S, et al. Shortened telomeres in circulating leukocytes of patients with chronic obstructive pulmonary disease. Am J Respir Crit Care Med 2009;179:566–571.
5. Morla M, Busquets X, Pons J, Sauleda J, MacNee W, Agusti AG. Telomere shortening in smokers with and without COPD. Eur Respir J 2006;27:525–528.
6. Tsuji T, Aoshiba K, Nagai A. Alveolar cell senescence in patients with pulmonary emphysema. Am J Respir Crit Care Med 2006;174:886–893.
7. Nyunoya T, Monick MM, Klingelhutz A, Yarovinsky TO, Cagley JR, Hunninghake GW. Cigarette smoke induces cellular senescence. Am J Respir Cell Mol Biol 2006;35:681–688.
8. Tsuji T, Aoshiba K, Nagai A. Cigarette smoke induces senescence in alveolar epithelial cells. Am J Respir Cell Mol Biol 2004;31:643–649.
9. Amsellem V, Gary-Bobo G, Marcos E, et al. Telomere dysfunction causes sustained inflammation in chronic obstructive pulmonary disease. Am J Respir Crit Care Med 2011;184:1358–1366.
10. Noureddine H, Gary-Bobo G, Alifano M, et al. Pulmonary artery smooth muscle cell senescence is a pathogenic mechanism for pulmonary hypertension in chronic lung disease. Circ Res 2011;109: 543–553.
11. Barnes PJ, Celli BR. Systemic manifestations and comorbidities of COPD. Eur Respir J 2009;33: 1165–1185.
12. Campisi J. Senescent cells, tumor suppression, and organismal aging: Good citizens, bad neighbors. Cell 2005;120:513–522.
13. Collado M, Blasco MA, Serrano M. Cellular senescence in cancer and aging. Cell 2007;130:223–233.
14. Matthews C, Gorenne I, Scott S, et al. Vascular smooth muscle cells undergo telomere-based senescence in human atherosclerosis: Effects of telomerase and oxidative stress. Circ Res 2006;99:156–164.
15. Sin DD, Man SF. Chronic obstructive pulmonary disease as a risk factor for cardiovascular morbidity and mortality. Proc Am Thorac Soc 2005;2:8–11.

16. Agusti A, Edwards LD, Rennard SI, et al. Persistent systemic inflammation is associated with poor clinical outcomes in COPD: A novel phenotype. PLoS One 2012;7:e37483.

17. Celli BR, Locantore N, Yates J, et al. Inflammatory biomarkers improve clinical prediction of mortality in chronic obstructive pulmonary disease. Am J Respir Crit Care Med 2012;185:1065–1072.

18. Yoshida T, Tuder RM. Pathobiology of cigarette smoke-induced chronic obstructive pulmonary disease. Physiol Rev 2007;87:1047–1082.

19. Sabit R, Bolton CE, Edwards PH, et al. Arterial stiffness and osteoporosis in chronic obstructive pulmonary disease. Am J Respir Crit Care Med 2007;175:1259–1265.

20. Bolton CE, Ionescu AA, Shiels KM, et al. Associated loss of fat-free mass and bone mineral density in chronic obstructive pulmonary disease. Am J Respir Crit Care Med 2004;170:1286–1293.

21. Coppe JP, Desprez PY, Krtolica A, Campisi J. The senescence-associated secretory phenotype: The dark side of tumor suppression. Annu Rev Pathol 2009;5:99–118.

22. Kuilman T, Peeper DS. Senescence-messaging secretome: SMS-ing cellular stress. Nat Rev Cancer 2009;9:81–94.

23. Kuilman T, Michaloglou C, Vredeveld LC, et al. Oncogene-induced senescence relayed by an interleukin-dependent inflammatory network. Cell 2008;133:1019–1031.

24. Coppe JP, Kauser K, Campisi J, Beausejour CM. Secretion of vascular endothelial growth factor by primary human fibroblasts at senescence. J Biol Chem 2006;281:29568–29574.

25. Acosta JC, O'Loghlen A, Banito A, et al. Chemokine signaling via the CXCR2 receptor reinforces senescence. Cell 2008;133:1006–1018.

26. Aoshiba K, Nagai A. Senescence hypothesis for the pathogenetic mechanism of chronic obstructive pulmonary disease. Proc Am Thorac Soc 2009;6:596–601.

27. Alder JK, Guo N, Kembou F, et al. Telomere length is a determinant of emphysema susceptibility. Am J Respir Crit Care Med 2011;184:904–912.

28. Yao H, Chung S, Hwang JW, et al. SIRT1 protects against emphysema via FOXO3-mediated reduction of premature senescence in mice. J Clin Invest 2012;122:2032–2045.

29. Nyunoya T, Monick MM, Klingelhutz AL, et al. Cigarette smoke induces cellular senescence via Werner's syndrome protein down-regulation. Am J Respir Crit Care Med 2009;179:279–287.

30. Alder JK, Chen JJ, Lancaster L, et al. Short telomeres are a risk factor for idiopathic pulmonary fibrosis. Proc Natl Acad Sci USA 2008;105:13051–13056.

31. Diaz de Leon A, Cronkhite JT, Katzenstein AL, et al. Telomere lengths, pulmonary fibrosis and telomerase (tert) mutations. PLoS One 2010;5:e10680.

32. Yoshida T, Mett I, Bhunia AK, et al. Rtp801, a suppressor of mTOR signaling, is an essential mediator of cigarette smoke-induced pulmonary injury and emphysema. Nat Med 2010;16:767–773.

33. Chaouat A, Bugnet AS, Kadaoui N, et al. Severe pulmonary hypertension and chronic obstructive pulmonary disease. Am J Respir Crit Care Med 2005;172:189–194.

34. Minamino T, Miyauchi H, Yoshida T, Ishida Y, Yoshida H, Komuro I. Endothelial cell senescence in human atherosclerosis: Role of telomere in endothelial dysfunction. Circulation 2002;105:1541–1544.

35. Murry CE, Gipaya CT, Bartosek T, Benditt EP, Schwartz SM. Monoclonality of smooth muscle cells in human atherosclerosis. Am J Pathol 1997;151:697–705.

36. Sanchez O, Marcos E, Perros F, et al. Role of endothelium-derived CC chemokine ligand 2 in idiopathic pulmonary arterial hypertension. Am J Respir Crit Care Med 2007;176:1041–1047.

37. Chaouat A, Savale L, Chouaid C, et al. Role for interleukin-6 in COPD-related pulmonary hypertension. Chest 2009;136:678–687.

38. Soon E, Holmes AM, Treacy CM, et al. Elevated levels of inflammatory cytokines predict survival in idiopathic and familial pulmonary arterial hypertension. Circulation 2010;122:920–927.

39. Krtolica A, Parrinello S, Lockett S, Desprez PY, Campisi J. Senescent fibroblasts promote epithelial cell growth and tumorigenesis: A link between cancer and aging. Proc Natl Acad Sci USA 2001;98:12072–12077.

40. Langen RC, Schols AM, Kelders MC, van der Velden JL, Wouters EF, Janssen-Heininger YM. Muscle wasting and impaired muscle regeneration in a murine model of chronic pulmonary inflammation. Am J Respir Cell Mol Biol 2006;35:689–696.

41. Marquis K, Debigare R, Lacasse Y, et al. Midthigh muscle cross-sectional area is a better predictor of mortality than body mass index in patients with chronic obstructive pulmonary disease. Am J Respir Crit Care Med 2002;166:809–813.

42. Swallow EB, Reyes D, Hopkinson NS, et al. Quadriceps strength predicts mortality in patients with moderate to severe chronic obstructive pulmonary disease. Thorax 2007;62:115–120.

43. Ferguson GT, Calverley PM, Anderson JA, et al. Prevalence and progression of osteoporosis in patients with COPD: Results from the towards a revolution in COPD health study. Chest 2009;136:1456–1465.

44. Mills NL, Miller JJ, Anand A, et al. Increased arterial stiffness in patients with chronic obstructive pulmonary disease: A mechanism for increased cardiovascular risk. Thorax 2008;63:306–311.

45. Willum-Hansen T, Staessen JA, Torp-Pedersen C, et al. Prognostic value of aortic pulse wave velocity as index of arterial stiffness in the general population. Circulation 2006;113:664–670.

46. Dransfield MT, Huang F, Nath H, Singh SP, Bailey WC, Washko GR. CT emphysema predicts thoracic aortic calcification in smokers with and without COPD. COPD 2010;7:404–410.

47. McAllister DA, MacNee W, Duprez D, et al. Pulmonary function is associated with distal aortic calcium, not proximal aortic distensibility. MESA lung study. COPD 2011;8:71–78.

48. Chandra D, Stamm JA, Palevsky PM, et al. The relationship between pulmonary emphysema and kidney function in smokers. Chest 2012;142:655–662.

49. Mannino DM, Thorn D, Swensen A, Holguin F. Prevalence and outcomes of diabetes, hypertension and cardiovascular disease in COPD. Eur Respir J 2008;32:962–969.

50. Kadunce DP, Burr R, Gress R, Kanner R, Lyon JL, Zone JJ. Cigarette smoking: Risk factor for premature facial wrinkling. Ann Intern Med 1991;114:840–844.

51. Patel BD, Loo WJ, Tasker AD, et al. Smoking related COPD and facial wrinkling: Is there a common susceptibility? Thorax 2006;61:568–571.

52. Bon J, Fuhrman CR, Weissfeld JL, et al. Radiographic emphysema predicts low bone mineral density in a tobacco-exposed cohort. Am J Respir Crit Care Med 2011;183:885–890.

53. McAllister DA, Maclay JD, Mills NL, et al. Arterial stiffness is independently associated with emphysema severity in patients with chronic obstructive pulmonary disease. Am J Respir Crit Care Med 2007;176:1208–1214.

54. Maclay JD, McAllister DA, Rabinovich R, et al. Systemic elastin degradation in chronic obstructive pulmonary disease. Thorax 2012;67:606–612.

55. Bernardes de Jesus B, Schneeberger K, Vera E, Tejera A, Harley CB, Blasco MA. The telomerase activator TA-65 elongates short telomeres and increases health span of adult/old mice without increasing cancer incidence. Aging Cell 2011;10:604–621.

56. Varela I, Pereira S, Ugalde AP, et al. Combined treatment with statins and aminobisphosphonates extends longevity in a mouse model of human premature aging. Nat Med 2008;14:767–772.

14 Lung Infections and Aging

Jacqueline M. Kruser and Keith C. Meyer

Department of Medicine, University of Wisconsin School of Medicine and Public Health, Madison, Wisconsin, USA

14.1 Introduction

By the year 2050, it is estimated that 20% of the United States population will be older than age 65, an increase from 12% in the year 2000 [1]. Lower respiratory tract infections continue to be a leading cause of morbidity and mortality in this population of aging adults. The annual incidence of pneumonia in community-dwelling elderly ranges from 25 to 44 cases per 1000 people, and in patients who live in nursing homes and other residential institutions, the incidence ranges from 33 to 114 per 1000 people [2]. The rates of hospitalization for pneumonia in the elderly range from 26 per 1000 people aged 75–84 up to 51 per 1000 people over the age of 85 [3]. Lower respiratory tract infections are not only common in the aging population but are also a significant cause of mortality. The risk of death during a hospital stay for pneumonia in elderly patients is 1.5 times higher when compared to the 10 other most frequent causes of hospitalization in this age group [3]. Infections with influenza and pneumonia were the seventh most common cause of death in patients aged 65 and older in 2007 and accounted for approximately 3% of all deaths in this age group [1], and tuberculosis remains as a major cause of morbidity and mortality worldwide [4]. Elderly patients hospitalized for pneumonia also have higher rates of morbidity and mortality related to cardiovascular disease following an acute respiratory infection [5]. Although pneumonia remains a leading cause of death, it is encouraging to note that the number of deaths attributed to influenza and pneumonia has decreased approximately 26% since 1999 [1].

14.2 Aging and immunosenescence

Multiple age-related changes contribute to the risk of pulmonary infections in the elderly, and age-associated immune system changes (immunosenescence) are undoubtedly an important component of this susceptibility (for a summary of known age-related changes, see Table 14.1). While many changes due to age-associated immune system remodeling may

Molecular Aspects of Aging: Understanding Lung Aging, First Edition. Edited by Mauricio Rojas, Silke Meiners and Claude Jourdan Le Saux.

Table 14.1 Age-associated changes that may increase susceptibility of the elderly to lung infections and dysregulated inflammatory responses.

Adaptive immunity.
- T cell function
  - Altered memory T cell function
    - ↑ Peripheral memory T cells
    - ↓ Proliferative responses
  - ↓ Naïve T lymphocyte production
  - ↓ CD8+ T cell cytotoxic responses
  - ↑ Treg number
  - Th1 to Th2 cytokine shift
  - ↑ HLA-DR expression
  - ↓ Treg target cell suppression and IL-10 production
  - ↑ CD4+CD28- T cells
  - ↑ IL-17-producing T cells
- B cell function
  - ↓ B cell number
  - ↓ Germinal center formation
  - ↓ Naïve and memory B cells
  - ↓ B cells in peripheral blood
  - Altered antibody responses to specific antigen
    - ↓ B lymphocyte receptor repertoire
    - ↓ Class switch recombination
    - ↓ Generation of high-affinity antibodies
    - ↑ IgA and IgG
    - ↑ Autoantibodies

Innate immunity
- ↓ Dendritic cell number and function
- Altered TLR function
- Dysregulated cytokine production
- ↑ Proinflammatory cytokines
- ↑ Circulating cytokines (e.g., IL-6, TNF-α)
- ↓ Macrophage and neutrophil function
- ↓ γδ T cell proliferation and number
- ↑ IL-17-producing NKT cells

Other changes
- ↑ Oxidant stress due to ROIs
- Telomere shortening; telomerase deficiency
- ↑ Stress kinase and transcription factor activation
- ↓ Antiaging molecule expression
- ↓ Glucocorticoid sensitivity
- ↓ Stem cell responses
- ↑ Proinflammatory cytokine production by senescent epithelial cells

HLA-DR, major histocompatibility complex DR class II protein; Ig, immunoglobulin; IL, interleukin; NKT, natural killer T; Th, helper T cell; TNF, tumor necrosis factor; TLR, Toll-like receptor; Treg, regulatory T cell.

result in a decline in immune function, it is important to note that complex interactions between the innate and adaptive arms of the immune system occur, and despite a general decline, some functions may be preserved or even more robust in the elderly [6, 7, 8]. The innate system is perceived as an ancient, nonspecific immune response that is highly conserved across most multicellular organisms [9]. The adaptive response is considered to be a more recent evolutionary development, and it is characterized by an antigen-specific response that involves pathogen recognition and memory [7, 8]. It is well recognized that adaptive immunity declines with aging as manifested by its waning ability to mount an antigen-specific response to novel pathogens, but there continues to be debate surrounding the innate immune system and whether aging leads to preserved, upregulated, or diminished innate function [6, 10, 11, 12]. These investigations suggest that innate immune responses may be either attenuated or overexuberant (or both) as a consequence of the immunosenescent changes that usually accompany advanced age.

14.2.1   Innate immunity

The cellular components of the innate immune system include neutrophils, monocytes, macrophages, dendritic cells, basophils, mast cells, eosinophils, and natural killer (NK) cells. These cell types are responsible for phagocytosis, the recognition of potential pathogens via pattern recognition receptors (PRRs) that recognize highly conserved molecular patterns on various classes of microorganisms that are termed pathogen-associated molecular patterns (PAMPs), the production of inflammatory mediators, the activation and modulation of the adaptive immune response, the recognition of injured self-tissues via damage-associated molecular patterns (DAMPs), and cell destruction. Noncellular components of the innate immune system (e.g., complement, acute-phase proteins, and cytokines) are also responsible for host defense at various levels including targeted cell death, relaying messages, and repair of damaged tissue [7]. The immediate response to exogenous stimuli by the innate immune system is mediated through PRRs, which include the Toll-like receptor (TLR) family [13]. These TLRs bind lipopolysaccharides, lipoteichoic acids, mannans, and bacterial DNA that are produced by pathogens, and this leads to the release of inflammatory mediators such as cytokines and costimulatory molecules [7]. In addition to the TLR family of receptors, two other families of PRRs have been described: the nucleotide-binding oligomerization domain (NOD)-like receptors (NLRs) and the RIG-like receptors (RLRs). These PRRs also appear to play important roles in innate immune responses [14]. While the innate immune system can initiate a strong inflammatory response independent of adaptive immunity, recruitment of both systems facilitates a more robust defense, and dendritic cells play a key role in the interactions between the two systems. After endocytosis of foreign antigens in the periphery, dendritic cells act as antigen-presenting cells that can travel to lymph nodes where they activate the T cell arm of the adaptive system [9].

In the lung, specialized components of the innate immune system create the first-line defense against the continuous exposure to aspirated or airborne pathogens. Antimicrobial peptides (including defensins, cathelicidins, and collectins) are secreted into the airways by respiratory tract cells to provide a first layer of defense through either direct antimicrobial properties or the promotion of phagocytosis [15]. Alveolar macrophages (AM), which make up the vast majority of airspace leukocytes that are responsible for the phagocytosis of both inert molecules and pathogenic organism, also produce proinflammatory cytokines, carry antigens to dendritic cells, and help recruit other immune cells to the lung [16]. Interestingly, AM appear to be capable of inducing T cell tolerance [17], which may help prevent overexuberant responses to lower tract infection.

Although the innate system is generally perceived as being better preserved than adaptive immunity in the aging adult, individual components of the innate immunity may undergo significant change with aging, and these changes may vary significantly among elderly individuals. It has been shown that a subset of older adults who were more than 85 years of age and demonstrated an impaired innate immune response characterized by decreased production of proinflammatory cytokines had a twofold increase in overall mortality, even when controlled for the presence of chronic illness [18]. Age-related decline in the innate system has been associated with decreasing phagocytic capacity and oxidative burst in neutrophils and macrophages, decreased cytotoxicity of NK cells, and decreased ability of dendritic cells to home to lymph nodes and stimulate T cells in *in vivo* studies [12]. However, components of the innate system may display significant augmentation with aging, such as an increased number of NK cells and increased serum levels of proinflammatory cytokines [10].

Age-associated changes in TLR responses are of increasing interest. Plasmacytoid dendritic cells (pDC) have been demonstrated to play a key role in the induction of innate immune responses following TLR9 triggering by virus in pDC-deficient mice [19], and this defect was associated with impaired recruitment of NK cells, neutrophils, monocytes, and macrophages to the site of inflammation. Interestingly, a decline in macrophage and pDC TLR responses has been reported with advanced age [20, 21], and TLRs are also expressed on other cell types (e.g., monocytes, neutrophils, B cells, epithelial cells, endothelial cells). Increased TLR4 expression has been described for lipid rafts as well as nonraft membrane functions for neutrophils in the basal state obtained from old individuals [22, 23]. Additionally, although dendritic cell TLR function appears to remain fairly intact with advancing age, an aging-associated decline in TLR1 and TLR4 expression by human monocytes has been reported, while TLR5 expression by peripheral blood monocytes has been shown to increase [24].

14.2.2 Adaptive immunity

Adaptive immunity is a more targeted response characterized by specificity for distinct pathogens and retained memory elicited by prior responses. The effector cells of the adaptive system (T and B lymphocytes) develop in the fetal liver and the bone marrow and then migrate to lymphoid tissues such as the spleen, lymph nodes, and mucosa-associated lymphoid tissue. T cell specificity is generated by the T cell receptor (TCR), which initiates maturation from naïve cells into antigen-specific effector and memory cells. B cell specificity is mediated through surface antibody receptors and secreted, soluble receptors.

Age-related changes in T cell function have been studied fairly extensively and are felt to play a leading role in immunosenescence. The thymus gland, which is necessary for the maturation of precursor hematopoietic cells into T cells, begins to involute in adolescence, and a majority of the functional gland is gradually replaced by fat by the age of 50 [12], although individuals typically have some remaining thymic tissues until age 60 [25]. As a result of this structural change, the number of naïve T cells produced by the thymus is significantly reduced in the elderly. In addition, naïve T cells in the elderly have a reduced functional response when stimulated by novel pathogens [26]. With aging, the total number of T cells remains constant, but there is a significant shift from populations of naïve T cells to a predominance of memory T cells, and the ability to mount an immune response to novel pathogens can become impaired [25], such as those responsible for lower respiratory tract infections. It has been proposed that this transition to memory T cells is a reflection of lifelong exposure to multiple and variable antigens, which leads to the replacement of naïve cells with specific memory T cells in the elderly [11].

Advancing age has been associated with a decline in TCR signaling, T cell proliferation in response to various stimuli, cognate helper function, cytokine production, and vaccine responses [27, 28], and one explanation for the decline in maximal CD4+ T cell proliferation responses in elderly individuals has been suggested to be telomere shortening [29, 30]. CD8+ T cells may also display declining function with advanced age. Impaired CD8+ T cell cytotoxicity responses have been associated with a waning ability to mount an adequate response to viral infections [31, 32, 33], and cell proliferation and IL-2 production by virus-specific memory CD8+ T cells have been shown to be impaired in aged mice [34]. Other observations include evidence of oligoclonal expansions of CD8+ T cells, which have been suggested to occur as a consequence of repeated exposures to microbial antigens, and it has been suggested that the accumulation of such clones may lead to less robust responses when new antigens are encountered [35].

Similarly, it has been shown in murine models of aging that the population of B lymphocytes also shifts from antigen-inexperienced B cells in younger individuals to activated, specific B cells in old mice as a result of chronic, lifelong antigen stimulation [36]. The immunoglobulin response of aging B cells is also diminished, with lower affinity for antigens [37] and lower antibody titers, and this is likely a major cause of the decreased response to influenza vaccination in elderly adults [38]. Class switch recombination becomes impaired with advancing age such that the ability to generate antibodies with the same specificity but different effector functions declines [39, 40], and a significant reduction in B cell repertoire diversity and switch memory B cells has been reported [41]. Immunoglobulin M (IgM) memory B cell function in response to pneumococcal vaccine appears to decline in the elderly [42], and lymphoreticular tissue mass in the nasopharynx or gut has been shown to decline with advanced age [43].

14.2.3 Autoimmunity

Autoimmune reactivity with erosion of self-tolerance tends to emerge with advanced age and may lead to impairment in both innate and adaptive immune mechanisms and promote the emergence of systemic autoimmune disease [44, 45]. Additionally, increased inflammation as a function of advanced age may impair the clearance of infectious agents as well as increase exposure to self-antigens [46, 47]. Finally, infections may trigger inflammatory disorders via mechanisms involving cross-reactivity of pathogen-associated epitopes with self or by disrupting signaling processes that maintain immune homeostasis [47, 48].

Th17 cells (a subset of CD4+ T cells) and the cytokine, IL-17, a potent proinflammatory mediator, have been identified as important regulators of immunity that are also linked to chronic inflammatory diseases and autoimmunity [49, 50], and aging appears to predispose to augmentation of IL-17 T cell responses [51, 52]. Although Th17 cells appear to play important roles in clearing extracellular pathogens such as *Klebsiella* spp., heightened IL-17 responses in aged mice can lead to increased tissue and organ inflammation (e.g., in response to viral infection) that can synergize with an impaired ability to contain virus (e.g., due to attenuated interferon-gamma (IFN-γ) production) and predispose elderly animals to a fatal outcome [52]. Additionally, an aging-associated decline in regulatory T cell (Treg) function may lead to augmented inflammation in response to a stimulus and/or autoimmunity due to loss of IL-17 restraint with advanced age [53]. Heightened Th17 and IL-17 responses as a consequence of advanced age may prolong inflammatory responses following an episode of community-acquired pneumonia (CAP) in the elderly [54], and IL-17-mediated inflammation may promote an increased degree of tissue damage, an increased likelihood of subsequent severe bacterial pneumonia, and an increased mortality risk with a viral infection (e.g., influenza) [55].

14.2.4 Lung-specific changes in immunity with aging

Although immunosenescence is generally described as a systemic process, there are important aging-associated changes in the immune response specific to the lungs, which have tremendous epithelial surface area and potential continual antigen exposure throughout life. Studies of bronchoalveolar lavage (BAL) fluid in elderly adults have demonstrated important age-related changes in both adaptive and innate immunity cellular components. Specifically, BAL fluid has been shown to have increased numbers of neutrophils and total lymphocytes, an increased CD4+/CD8+ T cell ratio, decreased number of B cells, increased IgM and IgA antibodies, as well as a decrease in vascular endothelial growth factor [56, 57, 58, 59]. Alveolar Macrophages (AM) are a critical aspect of lung-specific immunity, and multiple studies have demonstrated important functional changes in AM function with aging. In rat models that have examined macrophage responses, aging has been associated with decreased expression of TLRs [60], decreased generation of nitric oxide [61], and decreased TNF-α release [62]. Although human studies have not been able to replicate all of these results, Zissel and colleagues have demonstrated decreased accessory function in AM of elderly patients, which is necessary for T cell stimulation and activation of the adaptive response [63].

14.3 Inflamm-aging and susceptibility to infection

While multiple studies have described immunosenescence in elderly adults, aging and immune system remodeling has also been associated with changes that are suggestive of hyperinflammation, also known as **inflamm-aging**. Elderly patients have overall increased levels of circulating inflammatory mediators compared to younger cohorts (including proinflammatory cytokines such as IL-6 and TNF-α) as well as anti-inflammatory mediators such as acute-phase proteins and IL-1 receptor antagonist [64]. Franceschi and colleagues have hypothesized that inflamm-aging is a result of chronic, low-level antigen exposure throughout life and that macrophages play a central role in this process [65]. Macrophages play a critical part in first-line antigen processing and the activation of multiple other components of the inflammatory response. In this model, successful aging is characterized by a balance between production of proinflammatory agents and anti-inflammatory agents, while unsuccessful aging occurs when proinflammatory mediators increase without appropriate anti-inflammatory opposition [66].

Multiple cohort studies have demonstrated that increased levels of proinflammatory mediators in elderly populations are strong predictors of all-cause mortality [64], and so the clinical significance of inflamm-aging and the role of inflammatory mediators in specific age-related disorders such as cardiovascular disease, Alzheimer's disease, and diabetes has become an active area of investigation. In solid organ transplantation, for example, age-associated heightened inflammatory responses may play an important role in graft rejection; a murine model has shown that IL-17 levels increase with advancing age, and treatment with anti-IL-17 antibodies can delay acute allograft rejection [51]. Additionally, in a mouse model of viral infection, the treatment of aged mice with anti-IL-17 has been shown to prevent death by inhibiting neutrophil-mediated tissue necrosis [67].

Chronic obstructive pulmonary disease (COPD) is an age-related, chronic inflammatory lung disease that is characterized by increased numbers of inflammatory cells, increased levels of proinflammatory cytokines, and high levels of reactive oxygen intermediates (ROIs), and the senescent lung displays many of the same inflammation-mediated features with normal aging [68]. For details on the age-related changes seen in COPD, see Chapter 13. With both normal aging and COPD, patients experience an age-related decrease in their forced expiratory volume in 1s (FEV1), but the decline in lung function with normal aging is less than the change in patients who develop COPD [69], and alveolar wall destruction is not seen in normal lung senescence [70].

Free radicals, and specifically ROIs, are known to cause long-term tissue damage and play a significant role in the age-related changes of many organs [71] including the lung in both normal aging and COPD [68]. The accumulation of ROIs over time causes chronic oxidative stress that is known to damage DNA [72], which contributes to the programmed aging process. Senescent epithelial cells that have been damaged by ROIs can lose their ability to divide, but they can continue to remain active and produce high levels of proinflammatory molecules [54]. In addition, increasing levels of ROIs with age can activate transcription factors that lead to the further production of multiple proinflammatory molecules [73].

Inflamm-aging and its complex interactions with immunosenescence play an important role in the susceptibility of elderly adults to respiratory tract infections. Elderly individuals with increased baseline levels of proinflammatory mediators (IL-6 and TNF-α) have a higher risk of developing CAP [74], and patients with higher levels of IL-6 at the time of admission for pneumonia tend to have more severe infections [75]. Moreover, age-related cellular changes in the mouse lung related to increased inflammation have been shown to promote cellular adhesion of *Streptococcus pneumoniae* [76] and may lead to an increased risk of pneumococcal infection [77]. The ability of *S. pneumoniae* to bind to host cell surface protein ligands such as the polymeric immunoglobulin receptor (PIgR), platelet-activating factor receptor (PAFr), or the laminin receptor (LR) via binding of these ligands by bacterial cell wall or cell membrane constituents (phosphorylcholine-binding protein A, phosphorylcholine, or lipoteichoic acid) can increase 100-fold following *in vitro* pretreatment of host cells with TNF-α and IL-1β due to upregulation of the host cell surface ligands [78, 79]. Additionally, binding of pneumococci to LR and PAFr may be a key mechanism that facilitates invasion of pneumococci across the alveolar–capillary barrier, with resultant bacteremia [76, 80]. Aged mice have increased levels of these ligands versus younger mice [76], and increased expression of these ligands may allow bacteria to persist and replicate during early phases of lung infection. Additionally, aged mice appear to have reduced levels of TLR1, TLR2, and TLR4 in the lung [77], which may increase susceptibility to bacterial pneumonia.

In a study of healthy human volunteers, the induction of endotoxemia in the elderly was associated with a prolonged inflammatory response and a prolonged fever response when compared to younger subjects [81]. When compared to younger individuals, elderly patients hospitalized with pneumonia also have a prolonged *in vivo* inflammatory response to infection that is characterized by prolonged, high levels of proinflammatory mediators including TNF-α [82]. An improved understanding of the mechanisms of hyperinflammation and immune responses in the elderly may provide an opportunity to establish novel therapeutics to help prevent respiratory tract infections in this population. For example, Qian and colleagues have demonstrated that aging monocytes produce increased levels of TLR5 and IL-8, and they suggest that the elevated TLR5 activity may be targeted as a mechanism for enhancing the immune response to vaccination in the elderly [24].

14.4 Respiratory infection and regulation of host responses

When a host is challenged by a potential pathogen, a balanced immune response that limits pathogen proliferation and persistence yet avoids host-mediated collateral damage to the lung from an overexuberant inflammatory response is required. In addition to immune effector and memory cell functions, lymphocytes also differentiate into immunoregulatory cells. CD4+ T cells can differentiate into Tregs that can suppress or regulate inflammatory immune responses, and B cells can also differentiate into subsets with regulatory functions. The best described Treg-cell subset consists of CD4+ CD25+ lymphocytes that express the FOXP3 transcription factor, which is essential for this subset's regulatory function [45, 83, 84]. Additionally, regulatory B cells (Bregs) that have a number of functions, including regulation of T cell development and differentiation or suppression of cytokine production by monocytes, have been identified, and interleukin-10 (IL-10) plays a pivotal role in Breg function [85].

Pleuropulmonary infections are one of several manifestations of tuberculosis, and the immune response to infection with *Mycobacterium tuberculosis* has been studied extensively and will be used here to illustrate the balance between mounting a sufficient immune response to limit bacterial growth and dissemination yet avoiding excessive inflammation with consequent damage to host tissues. Both the innate and adaptive arms of the immune system must be appropriately activated for the host to mount an effective response to a mycobacterial infection. Mycobacteria activate several TLRs including TLR2, TLR4, and TLR9, and such stimulation leads to the augmentation of NK cell activity and triggers the induction of type 1 interferons (IFN) by lymphoid cells [86] as well as inducing the expression of multiple antimycobacterial effectors such as antimicrobial peptides that include cathelicidin [87], β-defensin [88], secretory leukocyte protease inhibitor [89], and lipocalin-2 [90]. Although several cell types can produce IFN-γ during the course of mycobacterial infection, the production of IFN-γ by CD4+ T cells must be induced and has been shown to play a key role during acute infection in a murine adoptive transfer model, and CD4+ T cell-derived IFN-γ was required to trigger a robust CD8+ T cell response [91].

However, in humans with active infection due to *M. tuberculosis*, it appears that *M. tuberculosis* can promote the elaboration of mediators such as IL-10 that downregulate and counteract CD4+ (Th1) and innate immune responses [92]. Additionally, the onset of induction of adaptive immune responses in humans is significantly delayed compared to that observed for other pathogens or various nonmicrobial antigens [93, 94], and such delay has also been observed in mouse models of mycobacterial infection [95, 96]. The causes of the delay in the onset of T cell responses have been attributed to the ability of *M. tuberculosis* to inhibit macrophage apoptosis (which delays acquisition of mycobacteria and mycobacterial antigens by dendritic cells) [97, 98] and to the early induction of Tregs [99, 100]. Increased susceptibility to mycobacterial infection in mice has also been linked to impaired recruitment of CD103+α(E) integrin-expressing dendritic cells at the onset of *M. tuberculosis* infection [101].

Experiments in mice have shown that Tregs expand preferentially in response to specific *M. tuberculosis* antigens, and adoptive transfer of specific Tregs increases bacterial load as priming of effector T cells and their arrival in the lung are delayed [102]. Tregs that are CD4+CD25+FoxP3+ have been isolated from peripheral blood of patients with active *M. tuberculosis* infection [103, 104], can suppress production of IFN-γ [104], and serve as a major source of IL-10 [105]. IL-10 is also produced by a variety of cells in addition to Tregs including various other T lymphocyte subsets (Th1, Th2, Th17), B cells, dendritic cell subsets, macrophages, and neutrophils. IL-10 is produced early in mycobacterial infection. Its inhibitory and anti-inflammatory properties may blunt both innate and adaptive responses to *M. tuberculosis* infection and promote long-term lung infection [106]. Of note, although pathogen-specific Tregs have been identified in a number of nonmycobacterial infections [107], specific Tregs have not been detected in some murine infections [108, 109], suggesting that pathogen-specific Tregs may not be induced by all infections.

Relatively little is known about the regulation of T cell responses during chronic infection with *M. tuberculosis*. IL-12 is required for naïve CD4+ T cells to differentiate into IFN-γ-producing effector cells, and sustained expression of IL-12 is required to maintain control of the infection [110, 111]. Effector T cells tend to be short-lived and undergo apoptosis following a period of intense, antigen-driven proliferation, and long-lived, antigen-specific T memory cells tend to be generated only after specific antigen has been cleared [112]. However, antigen-specific T cells appear to be maintained during persistent *M. tuberculosis* infection [113], and experiments suggest that an antigen-specific precursor population of CD4+ T cells are capable of maintaining a high degree of proliferative capacity and can give rise to a unique subset of terminally differentiated effector T cells that express the programmed death-1 (PD-1) cell surface marker [114]. PD-1 knockout mice show greatly increased susceptibility to *M. tuberculosis* infection with high bacterial burden in the lung and greater immunopathologic damage to lung tissues that is linked to a greatly increased expansion of antigen-specific CD4+ T cells [115]. If these CD4+ T cells are depleted, PD-1-deficient mice can be rescued from early death [116], suggesting that PD-1+ CD4+ T cells restrain the differentiation of antigen-specific CD4+ T cells, thereby preventing excessive, destructive lung inflammation yet maintaining persistent populations of antigen-specific T cells such that infection can be controlled.

Much more needs to be learned about the interplay of proinflammatory immune responses to infection and simultaneously initiated regulatory responses that serve to prevent overexuberant host responses that lead to excessive damage to tissues in which the infection occurs. Immunosenescent changes associated with the aging process may affect both the initial innate and adaptive immune responses to infection as well as the ability to mount an appropriate regulatory response, and more research is required to understand how such phenomena are affected by advancing/advanced age. The role of epigenetic factors in responses to infection [117] is a new area of investigation, and while responses to infection across species have many similarities, immune responses and pathogen-specific responses may vary significantly across species [118] such that animal modeling of respiratory infection, while providing important insights into disease pathogenesis, may not be identical to responses in humans. Additionally, it is conceivable that the gradual development of low-grade inflammation in the lung combined with a diminishing capacity for lung maintenance and repair (which may occur as a consequence of age-associated decline in the ability to suppress inflammation as anti-aging defenses wane) may play a significant role in the increased susceptibility of the elderly to develop lower respiratory tract infection. Additionally, exposure to particulates and pollutants in inhaled ambient air [119, 120], while possibly contributing to age-associated lung senescence and gradual loss of functioning, may also be of relevance for susceptibility to respiratory infection, and evidence has emerged that suggests that exposure to ambient air pollution increases the risk of developing CAP in the elderly [121, 122].

14.5 Preventing respiratory infection

Effective immunization may prevent respiratory tract infection for the elderly, and vaccines are currently available for both influenza and pneumococcal disease. Influenza vaccination rates in the United States are high, and 77% of adults over the age of 85 were vaccinated in 2008 [1]. However, the response to influenza vaccination is known to be less effective in the elderly when compared to younger adults because of declining immune system function [123]. Despite this diminished response, immunization against influenza is still somewhat effective in reducing the rates of hospitalization for pneumonia or influenza and reduces the overall risk of death in this population [124]. Interestingly, one study has shown that immunization against influenza in children is effective in preventing pneumonia and influenza in the elderly, possibly due to induction of herd immunity [125].

A 23-valent polysaccharide pneumococcal vaccine (PPV) is also in clinical use with the goal of decreasing the impact of respiratory tract infections due to pneumococcus in the elderly population as well as younger individuals with risk factors. Vaccination rates with PPV in the elderly were lower than influenza in 2008 with 54.6% of adults in the 65–74-year-old age group vaccinated and 68% of patients aged 75 and over [1]. A large systematic review of 22 studies has shown that pneumococcal vaccination is effective in preventing invasive pneumococcal disease (IPD), but vaccination did not substantially improve the incidence of all-cause pneumonia or reduce overall mortality [126]. Similar to the influenza vaccine, the PPV is also less effective in inducing an immune response in older adults compared to younger populations [127]. A study including only older adults, however, confirmed that the pneumococcal vaccine is effective in preventing IPD in this population and also demonstrated that the vaccine is effective in high-risk elderly populations [128]. According to CDC recommendations, all persons should be vaccinated with PPV once at age 65 years, and individuals who were vaccinated prior to age 65 for any reason (typically underlying comorbidity) should receive one repeat dose at or after age 65, with at least 5 years between doses [129]. Some groups have suggested, however, that more frequent vaccination schedules be considered, because studies have demonstrated that functional antibody levels can wane significantly by 5–10 years after vaccination, and revaccination in elderly adults is generally safe and effective in inducing a functional antibody response [130].

Attempts to make vaccination against respiratory pathogens more effective are ongoing. Vaccine effectiveness against influenza was estimated at only 47% against influenza A (H3N2) and 67% for influenza B infections for the 2012–2013 season in the United States ([131] Centers for Disease Control MMWR 2013), and effectiveness against influenza A in an elderly Danish population was only 11% [132]. Additionally, rapid decline in vaccine effectiveness against influenza A for the 2011–2012 seasons in Spain was observed mainly for elderly individuals (≥ age 65 years) [133]. Apart from the need for improved vaccines to prevent pneumococcal and viral pneumonia, an effective vaccine against *M. tuberculosis* would likely have a significant, worldwide effect on the morbidity and mortality associated with mycobacterial infection [134], and ongoing studies with novel vaccines [135] suggest that effective vaccination against tuberculosis may be possible via the induction of long-lived, specific CD4+ T cells.

Other strategies in addition to vaccination can also play an important role in preventing respiratory tract infections in the elderly. These include smoking cessation, minimizing aspiration risk, optimizing nutritional and functional status, adequate management of underlying chronic conditions, avoiding institutionalization, and judicious use of antimicrobial agents.

14.6 Summary and conclusions

Lower respiratory tract infections in the elderly are a significant source of morbidity and mortality. Age-related changes in both the systemic and the compartmentalized, lung-specific immune responses show characteristics of both functional decline (immunosenescence) and hyperinflammation (inflamm-aging), which are likely significant risk factors that increase the risk of respiratory tract infections in elderly adults. An improved understanding of aging-associated defects in immune responses that increase susceptibility to respiratory infection as well as mechanisms that prevent excessively exuberant responses to evolving respiratory tract infections that can cause irreversible lung damage is needed. Vaccines that can compensate for or overcome the blunted immune responses in the elderly are needed to maximize the impact of strategies that seek to prevent respiratory infection in the elderly.

References

1. National Center for Health Statistics. Health, United States, 2010. 2011; 1–563.
2. Janssens JP, Krause KH. Pneumonia in the very old. Lancet Infect Dis 2004; 4:112–124.
3. Fry AM, Shay DK, Holman RC, Curns AT, Anderson LJ. Trends in hospitalizations for pneumonia among persons aged 65 years or older in the United States, 1988–2002. JAMA 2005; 294:2712–2719.
4. Dye C. Global epidemiology of tuberculosis. Lancet 2006; 367(9514):938–940.
5. Perry TW, Pugh MJV, Waterer GW, et al. Incidence of cardiovascular events after hospital admission for pneumonia. Am J Med 2011; 124:244–251.
6. Franceschi C, Bonafè M. Centenarians as a model for healthy aging. Biochem Soc Trans 2003; 31: 457–461.
7. Delves PJ, Roitt IM. The immune system. First of two parts. N Engl J Med 2000; 343:37–49.
8. Delves PJ, Roitt IM. The immune system. Second of two parts. N Engl J Med 2000; 343:108–117.
9. Janeway CA Jr., Medzhitov R. Innate immune recognition. Annu Rev Immunol 2002; 20:197–216.
10. Gomez CR, Boehmer ED, Kovacs EJ. The aging innate immune system. Curr Opin Immunol 2005; 17:457–462.
11. Franceschi C, Bonafè M, Valensin S. Human immunosenescence: the prevailing of innate immunity, the failing of clonotypic immunity, and the filling of immunological space. Vaccine 2000; 18: 1717–1720.
12. Weiskopf D, Weinberger B, Grubeck-Loebenstein B. The aging of the immune system. Transpl Int 2009; 22:1041–1050.
13. Monie TP, Bryant CE, Gay NJ. Activating immunity: lessons from the TLRs and NLRs. Trends Biochem Sci 2009; 34:553–561.
14. Kawai T, Akira S. Toll-like receptors and their crosstalk with other innate receptors in infection and immunity. Immunity 2011; 34(5):637–650.
15. Zhang P, Summer WR, Bagby GJ, Nelson S. Innate immunity and pulmonary host defense. Immunol Rev 2000; 173:39–51.
16. Martin TR, Frevert CW. Innate immunity in the lungs. Proc Am Thorac Soc 2005; 2:403–411.
17. Coleman MM, Ruane D, Moran B, Dunne PJ, Keane J, Mills KH. Alveolar macrophages contribute to respiratory tolerance by inducing FoxP3 expression in naïve T cells. Am J Respir Cell Mol Biol 2013; 48:773–780.
18. van den Biggelaar AHJ, Huizinga TWJ, de Craen AJM, et al. Impaired innate immunity predicts frailty in old age. The Leiden 85-plus study. Exp Gerontol 2004; 39:1407–1414.
19. Guillerey C, Mouriès J, Polo G, et al. Pivotal role of plasmacytoid dendritic cells in inflammation and NK-cell responses after TLR9 triggering in mice. Blood 2012; 120(1):90–99.

20. Kovacs EJ, Palmer JL, Fortin CF, Fülöp T Jr., Goldstein DR, Linton PJ. Aging and innate immunity in the mouse: impact of intrinsic and extrinsic factors. Trends Immunol 2009; 30(7):319–324.

21. van Duin D, Shaw AC. Toll-like receptors in older adults. J Am Geriatr Soc 2007; 55(9):1438–1444.

22. Balistreri CR, Colonna-Romano G, Lio D, Candore G, Caruso C. TLR4 polymorphisms and ageing: implications for the pathophysiology of age-related diseases. J Clin Immunol 2009; 29(4):406–415.

23. Fulop T, Larbi A, Douziech N, et al. Signal transduction and functional changes in neutrophils with aging. Aging Cell 2004; 3(4):217–226.

24. Qian F, Wang X, Zhang L, et al. Age-associated elevation in TLR5 leads to increased inflammatory responses in the elderly. Aging Cell 2012; 11(1):104–110.

25. Linton PJ, Dorshkind K. Age-related changes in lymphocyte development and function. Nat Immunol 2004; 5:133–139.

26. Pfister G, Weiskopf D, Lazuardi L, et al. Naive T cells in the elderly: are they still there? Ann N Y Acad Sci 2006; 1067:152–157.

27. Lee N, Shin MS, Kang I. T-cell biology in aging, with a focus on lung disease. J Gerontol A Biol Sci Med Sci 2012; 67(3):254–263.

28. Dorshkind K, Montecino-Rodriguez E, Signer RA. The ageing immune system: is it ever too old to become young again? Nat Rev Immunol 2009; 9(1):57–62.

29. Weng NP, Hathcock KS, Hodes RJ. Regulation of telomere length and telomerase in T and B cells: a mechanism for maintaining replicative potential. Immunity 1998; 9(2):151–157.

30. Effros RB. Telomere/telomerase dynamics within the human immune system: effect of chronic infection and stress. Exp Gerontol 2011; 46(2–3):135–140.

31. Treanor J, Falsey A. Respiratory viral infections in the elderly. Antiviral Res 1999; 44(2):79–102.

32. Po JL, Gardner EM, Anaraki F, Katsikis PD, Murasko DM. Age-associated decrease in virus-specific CD8+ T lymphocytes during primary influenza infection. Mech Ageing Dev 2002; 123(8):1167–1181.

33. Blackman MA, Woodland DL. The narrowing of the CD8 T cell repertoire in old age. Curr Opin Immunol 2011; 23(4):537–542.

34. Effros RB, Walford RL. The immune response of aged mice to influenza: diminished T-cell proliferation, interleukin 2 production and cytotoxicity. Cell Immunol 1983; 81(2):298–305.

35. Clambey ET, Kappler JW, Marrack P. CD8 T cell clonal expansions & aging: a heterogeneous phenomenon with a common outcome. Exp Gerontol 2007; 42(5):407–411.

36. Johnson SA, Rozzo SJ, Cambier JC. Aging-dependent exclusion of antigen-inexperienced cells from the peripheral B cell repertoire. J Immunol 2002; 168:5014–5023.

37. Doria G, D'Agostaro G, Poretti A. Age-dependent variations of antibody avidity. Immunology 1978; 35:601.

38. Frasca D, Diaz A, Romero M, Landin AM, Blomberg BB. Age effects on B cells and humoral immunity in humans. Ageing Res Rev 2011; 10:330–335.

39. Frasca D, Blomberg BB. Aging affects human B cell responses. J Clin Immunol 2011; 31(3):430–435.

40. Lougaris V, Badolato R, Ferrari S, Plebani A. Hyper immunoglobulin M syndrome due to CD40 deficiency: clinical, molecular, and immunological features. Immunol Rev 2005; 203:48–66.

41. Gibson KL, Wu YC, Barnett Y, et al. B-cell diversity decreases in old age and is correlated with poor health status. Aging Cell 2009; 8(1):18–25.

42. Shi Y, Yamazaki T, Okubo Y, Uehara Y, Sugane K, Agematsu K. Regulation of aged humoral immune defense against pneumococcal bacteria by IgM memory B cell. J Immunol 2005; 175(5):3262–3267.

43. Fujihashi K, Kiyono H. Mucosal immunosenescence: new developments and vaccines to control infectious diseases. Trends Immunol 2009; 30(7):334–343.

44. Hasler P, Zouali M. Immune receptor signaling, aging, and autoimmunity. Cell Immunol 2005; 233:102–108.

45. Afzali B, Lombardi G, Lechler RI, Lord GM. The role of T helper 17 (Th17) and regulatory T cells (Treg) in human organ transplantation and autoimmune disease. Clin Exp Immunol 2007; 148:32–46.

46. Boren E, Gershwin ME. Inflamm-aging: autoimmunity, and the immune-risk phenotype. Autoimmun Rev 2004; 3:401–406.

47. Delogu LG, Deidda S, Delitala G, Manetti R. Infectious diseases and autoimmunity. J Infect Dev Ctries 2011; 5:679–687.

48. Mackay IR, Leskovsek NV, Rose NR. Cell damage and autoimmunity: a critical appraisal. J Autoimmun 2008; 30:5–11.

49. Miossec P. IL-17 and Th17 cells in human inflammatory diseases. Microbes Infect 2009; 11:625–630.

50. Maddur MS, Miossec P, Kaveri SV, Bayry J. Th17 cells: biology, pathogenesis of autoimmune and inflammatory diseases, and therapeutic strategies. Am J Pathol 2012; 181:8–18.

51. Tesar BM, Du W, Shirali AC, Walker WE, Shen H, Goldstein DR. Aging augments IL-17 T-cell alloimmune responses. Am J Transplant 2009; 9:54–63.

52. Goldstein DR. Aging, imbalanced inflammation and viral infection. Virulence 2010; 1:295.

53. Sun L, Hurez VJ, Thibodeaux SR, et al. Aged regulatory T cells protect from autoimmune inflammation despite reduced STAT3 activation and decreased constraint of IL-17 producing T cells. Aging Cell 2012; 11:509–519.

54. Boyd AR, Orihuela CJ. Dysregulated inflammation as a risk factor for pneumonia in the elderly. Aging Dis 2011; 2:487–500.

55. Goldstein DR. Role of aging on innate responses to viral infections. J Gerontol A Biol Sci Med Sci 2012; 67:242–246.

56. Meyer KC, Ershler W, Rosenthal NS, Lu XG, Peterson K. Immune dysregulation in the aging human lung. Am J Respir Crit Care Med 1996; 153:1072–1079.

57. Meyer KC, Rosenthal NS, Soergel P, Peterson K. Neutrophils and low-grade inflammation in the seemingly normal aging human lung. Mech Ageing Dev 1998; 104:169–181.

58. Meyer KC, Soergel P. Variation of bronchoalveolar lymphocyte phenotypes with age in the physiologically normal human lung. Thorax 1999; 54:697–700.

59. Thompson AB, Scholer SG, Daughton DM, Potter JF, Rennard SI. Altered epithelial lining fluid parameters in old normal individuals. J Gerontol 1992; 47:M171–M176.

60. Boehmer ED, Goral J, Faunce DE, Kovacs EJ. Age-dependent decrease in Toll-like receptor 4-mediated proinflammatory cytokine production and mitogen-activated protein kinase expression. J Leukoc Biol 2004; 75:342–349.

61. Koike E, Kobayashi T, Mochitate K, Murakami M. Effect of aging on nitric oxide production by rat alveolar macrophages. Exp Gerontol 1999; 34:889–894.

62. Corsini E, Battaini F, Lucchi L, et al. A defective protein kinase C anchoring system underlying age-associated impairment in TNF-alpha production in rat macrophages. J Immunol 1999; 163:3468–3473.

63. Zissel G, Schlaak M, Müller-Quernheim J. Age-related decrease in accessory cell function of human alveolar macrophages. J Investig Med 1999; 47:51–56.

64. Krabbe KS, Pedersen M, Bruunsgaard H. Inflammatory mediators in the elderly. Exp Gerontol 2004; 39:687–699.

65. Franceschi C, Bonafè M, Valensin S, et al. Inflamm-aging. An evolutionary perspective on immunosenescence. Ann N Y Acad Sci 2000; 908:244–254.

66. Franceschi C, Capri M, Monti D, et al. Inflammaging and anti-inflammaging: a systemic perspective on aging and longevity emerged from studies in humans. Mech Ageing Dev 2007; 128:92–105.

67. Stout-Delgado HW, Du W, Shirali AC, Booth CJ, Goldstein DR. Aging promotes neutrophil-induced mortality by augmenting IL-17 production during viral infection. Cell Host Microbe 2009; 6:446–456.

68. Ito K, Barnes PJ. COPD as a disease of accelerated lung aging. Chest 2009; 135:173–180.

69. Fletcher C, Peto R. The natural history of chronic airflow obstruction. BMJ 1977; 1:1645–1648.

70. Verbeken E, Cauberghs M, Mertens I, Clement J, Lauweryns J, Van de Woestijne K. The senile lung. Comparison with normal and emphysematous lungs. 1. Structural aspects. Chest 1992; 101:793–799.

71. Harman, D. Free radical theory of aging: an update. Ann N Y Acad Sci 2006; 1067:10–21.

72. Promislow DEL. DNA repair and the evolution of longevity: a critical analysis. J Theor Biol 1994; 170:291–300.

73. Rahman I, Adcock IM. Oxidative stress and redox regulation of lung inflammation in COPD. Eur Respir J 2006; 28:219–242.

74. Yende S, Tuomanen EI, Wunderink R, et al. Preinfection systemic inflammatory markers and risk of hospitalization due to pneumonia. Am J Respir Crit Care Med 2005; 172:1440–1446.

75. Antunes G, Evans SA, Lordan JL, Frew AJ. Systemic cytokine levels in community-acquired pneumonia and their association with disease severity. Eur Respir J 2002; 20:990–995.

76. Shivshankar P, Boyd AR, Le Saux CJ, Yeh IT, Orihuela CJ. Cellular senescence increases expression of bacterial ligands in the lungs and is positively correlated with increased susceptibility to pneumococcal pneumonia. Aging Cell 2011; 10:798–806.

77. Hinojosa E, Boyd AR, Orihuela CJ. Age-associated inflammation and toll-like receptor dysfunction prime the lungs for pneumococcal pneumonia. J Infect Dis 2009; 200(4):546–554.

78. Zhang JR, Mostov KE, Lamm ME, et al. The polymeric immunoglobulin receptor translocates pneumococci across human nasopharyngeal epithelial cells. Cell 2000; 102(6):827–837.

79. Cundell DR, Gerard NP, Gerard C, Idanpaan-Heikkila I, Tuomanen EI. Streptococcus pneumoniae anchor to activated human cells by the receptor for platelet-activating factor. Nature 1995; 377(6548):435–438.

80. Orihuela CJ, Mahdavi J, Thornton J, et al. Laminin receptor initiates bacterial contact with the blood brain barrier in experimental meningitis models. J Clin Invest 2009; 119(6):1638–1646.

81. Krabbe KS, Bruunsgaard H, Hansen CM, et al. Ageing is associated with a prolonged fever response in human endotoxemia. Clin Diagn Lab Immunol 2001; 8:333–338.

82. Bruunsgaard H, Skinhøj P, Qvist J, Pedersen BK. Elderly humans show prolonged in vivo inflammatory activity during pneumococcal infections. J Infect Dis 1999; 180:551–554.

83. Peterson RA. Regulatory T-cells: diverse phenotypes integral to immune homeostasis and suppression. Toxicol Pathol 2012; 40(2):186–204.

84. Goodman WA, Cooper KD, McCormick TS. Regulation generation: the suppressive functions of human regulatory T cells. Crit Rev Immunol 2012; 32(1):65–79.

85. Mauri C, Bosma A. Immune regulatory function of B cells. Annu Rev Immunol 2012; 30:221–241.

86. Saiga H, Shimada Y, Takeda K. Innate immune effectors in mycobacterial infection. Clin Dev Immunol. 2011; 2011:347594..

87. Liu PT, Stenger S, Tang DH, Modlin RL. Cutting edge: vitamin D-mediated human antimicrobial activity against Mycobacterium tuberculosis is dependent on the induction of cathelicidin. J Immunol 2007; 179(4):2060–2063.

88. Rivas-Santiago B, Schwander SK, Sarabia C, et al. Human {beta}-defensin 2 is expressed and associated with Mycobacterium tuberculosis during infection of human alveolar epithelial cells. Infect Immun 2005; 73(8):4505–4511.

89. Nishimura J, Saiga H, Sato S, et al. Potent antimycobacterial activity of mouse secretory leukocyte protease inhibitor. J Immunol 2008; 180(6):4032–4039.

90. Saiga H, Nishimura J, Kuwata H, et al. Lipocalin 2-dependent inhibition of mycobacterial growth in alveolar epithelium. J Immunol 2008; 181(12):8521–8527.

91. Green AM, Difazio R, Flynn JL. IFN-γ from CD4 T cells is essential for host survival and enhances CD8 T cell function during Mycobacterium tuberculosis infection. J Immunol 2013; 190(1): 270–277.

92. Almeida AS, Lago PM, Boechat N, et al. Tuberculosis is associated with a down-modulatory lung immune response that impairs Th1-type immunity. J Immunol 2009; 183(1):718–731.

93. Urdahl KB, Shafiani S, Ernst JD. Initiation and regulation of T-cell responses in tuberculosis. Mucosal Immunol 2011; 4(3):288–293.

94. Miller JD, van der Most RG, Akondy RS, et al. Human effector and memory CD8+ T cell responses to smallpox and yellow fever vaccines. Immunity 2008; 28(5):710–722.

95. Chackerian AA, Alt JM, Perera TV, Dascher CC, Behar SM. Dissemination of Mycobacterium tuberculosis is influenced by host factors and precedes the initiation of T-cell immunity. Infect Immun 2002; 70(8):4501–4509.

96. Gallegos AM, Pamer EG, Glickman MS. Delayed protection by ESAT-6-specific effector CD4+ T cells after airborne M. tuberculosis infection. J Exp Med 2008; 205(10):2359–2368.

97. Chen M, Divangahi M, Gan H, et al. Lipid mediators in innate immunity against tuberculosis: opposing roles of PGE2 and LXA4 in the induction of macrophage death. J Exp Med 2008; 205(12):2791–2801.

98. Divangahi M, Desjardins D, Nunes-Alves C, Remold HG, Behar SM. Eicosanoid pathways regulate adaptive immunity to *Mycobacterium tuberculosis*. Nat Immunol 2010; 11(8):751–758.

99. Kursar M, Koch M, Mittrücker HW, et al. Cutting edge: regulatory T cells prevent efficient clearance of *Mycobacterium tuberculosis*. J Immunol 2007; 178(5):2661–2665.

100. Scott-Browne JP, Shafiani S, Tucker-Heard G, et al. Expansion and function of Foxp3-expressing T regulatory cells during tuberculosis. J Exp Med 2007; 204(9):2159–2169.

101. Leepiyasakulchai C, Ignatowicz L, Pawlowski A, Källenius G, Sköld M. Failure to recruit anti-inflammatory CD103+ dendritic cells and a diminished CD4+ Foxp3+ regulatory T cell pool in mice that display excessive lung inflammation and increased susceptibility to Mycobacterium tuberculosis. Infect Immun 2012; 80(3):1128–1139.

102. Shafiani S, Tucker-Heard G, Kariyone A, Takatsu K, Urdahl KB. Pathogen-specific regulatory T cells delay the arrival of effector T cells in the lung during early tuberculosis. J Exp Med 2010; 207(7):1409–1420.

103. Hougardy JM, Verscheure V, Locht C, Mascart F. In vitro expansion of CD4+ CD25highFOXP3+CD127-low/- regulatory T cells from peripheral blood lymphocytes of healthy *Mycobacterium tuberculosis*-infected humans. Microbes Infect 2007; 9(11):1325–1332.

104. Chen X, Zhou B, Li M, et al. CD4(+)CD25(+)FoxP3(+) regulatory T cells suppress Mycobacterium tuberculosis immunity in patients with active disease. Clin Immunol 2007; 123(1):50–59.

105. O'Garra A, Vieira PL, Vieira P, Goldfeld AE. IL-10-producing and naturally occurring CD4+ Tregs: limiting collateral damage. J Clin Invest 2004; 114(10):1372–1378.

106. Redford PS, Murray PJ, O'Garra A. The role of IL-10 in immune regulation during *M. tuberculosis* infection. Mucosal Immunol 2011; 4(3):261–270.

107. Belkaid Y, Tarbell K. Regulatory T cells in the control of host-microorganism interactions (*). Annu Rev Immunol 2009; 27:551–589.

108. Ertelt JM, Rowe JH, Johanns TM, Lai JC, McLachlan JB, Way SS. Selective priming and expansion of antigen-specific Foxp3- CD4+ T cells during Listeria monocytogenes infection. J Immunol 2009; 182(5):3032–3038.

109. Antunes I, Tolaini M, Kissenpfennig A, et al. Retrovirus-specificity of regulatory T cells is neither present nor required in preventing retrovirus-induced bone marrow immune pathology. Immunity 2008; 29(5):782–794.

110. Cooper AM, Magram J, Ferrante J, Orme IM. Interleukin 12 (IL-12) is crucial to the development of protective immunity in mice intravenously infected with mycobacterium tuberculosis. J Exp Med 1997; 186(1):39–45.

111. Feng CG, Jankovic D, Kullberg M, et al. Maintenance of pulmonary Th1 effector function in chronic tuberculosis requires persistent IL-12 production. J Immunol 2005; 174(7):4185–4192.

112. Sallusto F, Lanzavecchia A, Araki K, Ahmed R. From vaccines to memory and back. Immunity 2010; 33(4):451–463.

113. Winslow GM, Roberts AD, Blackman MA, Woodland DL. Persistence and turnover of antigen-specific CD4 T cells during chronic tuberculosis infection in the mouse. J Immunol 2003; 170(4):2046–2052.

114. Reiley WW, Shafiani S, Wittmer ST, et al. Distinct functions of antigen-specific CD4 T cells during murine Mycobacterium tuberculosis infection. Proc Natl Acad Sci USA 2010; 107(45):19408–19413.

115. Lázár-Molnár E, Chen B, Sweeney KA, et al. Programmed death-1 (PD-1)-deficient mice are extraordinarily sensitive to tuberculosis. Proc Natl Acad Sci USA 2010; 107(30):13402–13407.

116. Barber DL, Mayer-Barber KD, Feng CG, Sharpe AH, Sher A. CD4 T cells promote rather than control tuberculosis in the absence of PD-1-mediated inhibition. J Immunol 2011; 186(3):1598–1607.

117. Schulte LN, Eulalio A, Mollenkopf HJ, Reinhardt R, Vogel J. Analysis of the host microRNA response to Salmonella uncovers the control of major cytokines by the let-7 family. EMBO J 2011; 30(10):1977–1989.

118. Zinman G, Brower-Sinning R, Emeche CH, et al. Large scale comparison of innate responses to viral and bacterial pathogens in mouse and macaque. PLoS One 2011; 6(7):e22401.

119. Hogg JC, van Eeden S. Pulmonary and systemic response to atmospheric pollution. Respirology 2009; 14(3):336–346.

120. Oberdörster G. Pulmonary effects of inhaled ultrafine particles. Int Arch Occup Environ Health 2001; 74(1):1–8.

121. Neupane B, Jerrett M, Burnett RT, Marrie T, Arain A, Loeb M. Long-term exposure to ambient air pollution and risk of hospitalization with community-acquired pneumonia in older adults. Am J Respir Crit Care Med 2010; 181(1):47–53.

122. Zanobetti A, Bind MA, Schwartz J. Particulate air pollution and survival in a COPD cohort. Environ Health 2008; 7:48.

123. Reber AJ, Chirkova T, Kim JH, et al. Immunosenescence and challenges of vaccination against influenza in the aging population. Aging Dis 2012; 3:68.

124. Nichol KL, Nordin JD, Nelson DB, Mullooly JP, Hak E. Effectiveness of influenza vaccine in the community-dwelling elderly. N Engl J Med 2007; 357:1373–1381.

125. Cohen SA, Chui KKH, Naumova EN. Influenza vaccination in young children reduces influenza-associated hospitalizations in older adults, 2002–2006. J Am Geriatr Soc 2011; 59:327–332.

126. Moberley SA, Holden J, Tatham DP, Andrews RM. Vaccines for preventing pneumococcal infection in adults. Cochrane Database Syst Rev 2008; 1:CD000422.

127. Westerink MJ, Schroeder HW, Nahm MH. Immune responses to pneumococcal vaccines in children and adults: rationale for age-specific vaccination. Aging Dis 2012; 3:51.

128. Vila-Corcoles A, Ochoa-Gondar O, Guzman JA, Rodriguez-Blanco T, Salsench E, Fuentes CM. Effectiveness of the 23-valent polysaccharide pneumococcal vaccine against invasive pneumococcal disease in people 60 years or older. BMC Infect Dis 2010; 10:73.

129. Centers for Disease Control and Prevention (CDC). Advisory Committee on Immunization Practices. Updated recommendations for prevention of invasive pneumococcal disease among adults using the 23-valent pneumococcal polysaccharide vaccine (PPSV23). MMWR Morb Mortal Wkly Rep 2010; 59:1102–1106.

130. Grabenstein JD, Manoff SB. Pneumococcal polysaccharide 23-valent vaccine: long-term persistence of circulating antibody and immunogenicity and safety after revaccination in adults. Vaccine 2012; 30:4435–4444.

131. Centers for Disease Control and Prevention (CDC). Interim adjusted estimates of seasonal influenza vaccine effectiveness – United States, February 2013. MMWR Morb Mortal Wkly Rep 2013; 62(7): 119–123.

132. Bragstad K, Emborg H, Kolsen Fischer T, et al. Low vaccine effectiveness against influenza A(H3N2) virus among elderly people in Denmark in 2012/13 – a rapid epidemiological and virological assessment. Euro Surveill 2013; 18(6): 11-17.

133. Castilla J, Martinez-Baz I, Martinez-Artola V, et al. Early estimates of influenza vaccine effectiveness in Navarre, Spain: 2012/13 mid-season analysis. Euro Surveill 2013; 18(7):2.

134. Zevallos M, Justman JE. Tuberculosis in the elderly. Clin Geriatr Med 2003; 19:121–138.

135. Day CL, Tameris M, Mansoor N, et al. Induction and regulation of T cell immunity by the novel TB vaccine M72/AS01 in South African adults. Am J Respir Crit Care Med 2013; 15:492–502.

Index

Molecular Aspects of Aging: Understanding Lung Aging, First Edition. Edited by Mauricio Rojas, Silke Meiners and Claude Jourdan Le Saux.
© 2014 John Wiley & Sons, Inc. Published 2014 by John Wiley & Sons, Inc.